STYLE, POLITICS AND THE FUTURE OF PHILOSOPHY

BOSTON STUDIES IN THE PHILOSOPHY OF SCIENCE

VOLUME 114

ALLAN JANIK
Brenner Archive
Innsbruck University

STYLE, POLITICS AND THE FUTURE OF PHILOSOPHY

KLUWER ACADEMIC PUBLISHERS
DORDRECHT / BOSTON / LONDON

Library of Congress Cataloging-in-Publication Data

Janik, Allan.
Style, politics, and the future of philosophy.

(Boston studies in the philosophy of science)
Includes index.
1. Wittgenstein, Ludwig, 1889-1951. 2. Political science--Philosophy. 3. Social sciences--Philosophy. I. Title. II. Series.
Q174.B67 [B3376.W564] 001'.01 s [190] 88-32580
ISBN 0-7923-0056-4 (U.S.)

Published by Kluwer Academic Publishers
P.O. Box 17, 3300 AA Dordrecht, The Netherlands

Kluwer Academic Publishers incorporates the publishing programmes of D. Reidel, Martinus Nijhoff, Dr W. Junk and MTP Press.

Sold and distributed in the U.S.A. and Canada by Kluwer Academic Publishers, 101 Philip Drive, Norwell, MA 02061, U.S.A.

In all other countries, sold and distributed by Kluwer Academic Publishers Group, P.O. Box 322, 3300 AH Dordrecht, The Netherlands

printed on acid free paper

Printed in the Netherlands

For Bosse, Dexi and Kjell

The first and last schoolmaster of life is living and committing oneself unreservedly and dangerously to living; to men who know this an Aristotle and a Plato have much to say; but those who have imposed cautions upon themselves and petrified themselves in a system of ideas, them the masters themselves will lead into error.

Thornton Wilder, *The Ides of March*

TABLE OF CONTENTS

	Preface	ix
I.	Style and Idea in the Later Heidegger: Rhetoric, Politics and Philosophy	1
II.	Nyíri on the Conservatism of Wittgenstein's Later Philosophy	40
III.	Wittgenstein, Marx and Sociology	59
IV.	On Edification and Cultural Conversation: A Critique of Rorty	80
V.	Towards a Wittgensteinian Metaphysics of the Political	93
VI.	Culture, Controversy and the Human Studies	109
VII.	The Politics of Conciliation	116
VIII.	Discussing Technology — Breaking the Ground	133
IX.	Socialization is Creative Because Creativity is Social	146
X.	Myth and Certainty	159
XI.	Self-Deception, Naturalism and Certainty: Prolegomena to a Critical Hermeneutics	172
XII.	Psychoanalysis: Science, Literature or Art?	190
XIII.	Between Enlightenment and Counter-Enlightenment: The Self-Critical Rationalism of G. C. Lichtenberg	197
XIV.	Tacit Knowledge, Working Life and Scientific Method	211
XV.	MARTIN BRODY and ALLAN JANIK / Paradigms, Politics and Persuasion: Sociological Aspects of Musical Controversy	225
	Afterword with Acknowledgements	264
	Index of Names	268

PREFACE

Why did the two most influential philosophers in the twentieth century, Ludwig Wittgenstein and Martin Heidegger, write in such a curious fashion that they confused a whole generation of disciples and created a cottage industry for a second generation in the interpretation of their works? Do those curious writing strategies have a philosophical significance? How does philosophical style reflect attitudes to society and politics or bear significance for the social sciences? Is politics one type of human activity among many other independent ones as the classical modern political theorists from Hobbes and Machiavelli onwards have thought, or is it part and parcel of all of the activities into which an animal that speaks enters? How could the latter be elucidated? If politics arises from legitimate disputes about meanings, what does this imply for current cultural debates? for the so-called social sciences? above all, for that cultural conversation which some consider to be the destiny of philosophy in the wake of the demise of foundationalism?

These are a few of the most important questions which led me to the critical confrontation and reflections in the essays collected below. Written over a ten year period, they arise out of longstanding pondering on a wide variety of issues: the one-sided and clichéd character of most standard accounts of twentieth century philosophy and its development (which persist despite the efforts of brilliant historians such as John Passmore), Aristotle's concept of *praxis*, the Marxian critique of philosophy, the relationship of style and idea in writers like Karl Kraus and Denis Diderot, the rejection of the problem of interpretation in so much of contemporary analytic philosophy, the question of what role philosophy should play in society. These essays represent the tip of an iceberg in the sense that they are the small part of my reflections that have found their way into explicit and systematic formulation in some instances as a result of invitations to lecture on a particular topic or as the result of encountering a thesis (as in the case of my encounters with Minder, Nyíri and Olsson), which elicited an immediate response in polemical form. A third set of essays are a kind of resumé of the sort of cooperative work between philosophers and, say, music theorists or working life researchers, which I take to be one very important role that philosophy can play in the future.

I have chosen to put my critique of Robert Minder's attack on Heidegger at the beginning of this collection because it exemplifies the ways in which the history of ideas' approach to the problem of style and idea can be genuinely illuminating at a philosophical level. It is hardly a secret that there is no case in which the relation of style in politics and in philosophy is more troubling than that of Heidegger: his involvement with National Socialism and the general obscurity of his thought clearly invite writing him off as an obscurantist or even worse to those who have been weaned on the values of clarity and distinctness. Nobody has denounced Heidegger with more learning or eloquence than the late French critic, Robert Minder. Yet, for all the learning and eloquence which informs Minder's analysis the question of the philosophical significance of Heidegger's path of thinking remains untouched. Thus, I have tried to show how it is possible to understand Heidegger's way of thinking without either reading it through his biography alone or, for that matter, endorsing his views.

The center of these essays, however, is the interpretation of Wittgenstein's later philosophy especially as it has implications for social thought. Like Heidegger, Wittgenstein is an astonishingly difficult thinker to approach as it were from without. No small part of the difficulties in approaching Wittgenstein have to do with his typical modes of expression, unanswered questions, analogies, aphorisms, and curious examples which generally, when contrasted with those, say, of Carnap or Quine often seem to the uninitiated as obscurantist as those of Heidegger. Thus some early infidels went so far as to brand him an outright charlatan; while others write books, not about what he said or how he could be criticized, but about what they think he said, which in the end only serve to make Wittgenstein yet more 'kriptic' by eliminating just what he took to be most important in his work. The idea that Wittgenstein was a philosophical primitive, the Douanier Rousseau of philosophy as it were, seems to motivate such desires to 'reconstruct' his thought: just as Douanier Rousseau's style was the result of his not being able to paint any other way (something often assumed but not universally accepted by the way), Wittgenstein wrote the way he did out of necessity rather than choice, capable of brilliant intuition but unable to express himself in clear and distinct arguments in the way that a philosopher should. Ben Tilghman has put the point nicely while writing about the view of Wittgenstein in a segment of the American philosophical community: "In this country too many people still think of him as a "linguistic" philosopher with some quirky theories about language and psychology that the work of Quine, Davidson et al. has shown to be misguided". Like Tilghman who has labored yet longer in the vineyards than I have, I find this view basically misguided. A quarter of century of research, beginning with a wholly skeptical attitude to what I first took to be a vastly inflated reputation, has led me to two very important dissenting conclusions. First, Wittgenstein's conception of philosophy was thought out in full awareness of the alternatives. Second, his style reflects the results of that remarkable confrontation with the western philosophical tradition from Socrates onwards according to which the philosopher's task has been transformed from resolving problems on the basis of a theory to that of dissolving them in tortuous reflections on the nature of practice, which can only be accomplished through a certain linguistic *gesturing*. A clear grasp of those practices is only possible by means of linguistic gesture precisely because our practices are constituted through language; and that very 'fact' implies that there is no point outside of them from which they can be viewed in the way that the Newtonian ideal observer views physical phenomena. A good part of the essays below are devoted to spelling this out in various ways.

It is central to the view that I develop here that Wittgenstein's philosophizing, like Heidegger's, is not determined by personal whim. It is rooted in dissatisfaction with traditional ways of philosophizing from Plato and Aristotle through Descartes and Hume to Husserl and Russell, as, say, Rorty present them. It is but a way of challenging longstanding "absolute presuppositions" in Collingwood's felicitous phrase and one which is required by the very nature of the redirecting of inquiry. Thus, despite all other surface similarities (and there are many), Wittgenstein and Heidegger remain essentially different from figures like John Dewey and Otto Neurath, whose view of philosophy is essentially tied to what they want society to become. This is fundamental to my criticism of Rorty. On a Wittgensteinian view of the matter it is precisely the social engagement of philosophers which preserves them from taking the crucial final step that sets the desire to philosophize in the traditional sense to rest once and for all. This step goes beyond rejecting traditional epistemology on the basis of its irrelevance to our actual scientific practices but seeks to remove the conundrums that arise as we try to grasp the nature of those practices theoretically by showing how the conditions that make them possible are embedded in the natural history of a language-using creature.

My aim in this part of the book has been to tease out a coherent heuristic for social thought from the central notion in Wittgenstein's later philosophy through a criticism of certain theses about the implications of Wittgenstein's philosophizing in those fields. Moreover, that aim has been to steer a course away from superficial or tangential readings of Wittgenstein in this context and to articulate the (often elusive) implications of his central insights concerning rule-following behavior for our understanding of society and politics. However, it is crucial to emphasize here that this has not been in aid of developing anything like a discipline, The Philosophy of Social Science, for as Wittgenstein might put it the social sciences must take care of themselves. The philosophical task has more to do with developing a set of hints or reminders concerning both the pitfalls which await social scientists and heuristics for the orientation of research.

Wittgenstein is crucial to this task on account of the practice-immanent notion of rationality which lies at the center of his thinking. In contrast to two main approaches to the matter, so-called Critical Theory and so-called Deconstructionism, Wittgenstein neither takes it that the task of philosophy *can be* to create, postulate or construct rational standards of conduct over against the existing ones; nor does he despair, given the lack of such a 'rational' morality, that there can be one. Rather, he suggests (at least as I read him) that we have been looking for morality in the wrong place. Instead he offers us a third way, which avoids the dilemma by suggesting that we should have been looking for the morality and the rationality that is immanent in our practices as opposed to seeking something over against those practices, which will provide us with principles for criticizing established behavior patterns with a view to transforming them. Once morality (the Rational) gets set over against politics (the Real) we put ourselves into a situation where either politics has to be considered basically irrational or morality becomes mere wishful thinking. But are these sorts of alternatives forced upon us? I doubt it. The crux of the matter has to do with what we make of disagreement: is it the result of a failure to achieve consensus or is it a ground for rejecting rationality entirely in favor of an "anything goes" attitude? More important, I suggest, is the notion that both critical theorists and deconstructionists pose the problem falsely, because speculatively. Critical theory fails to look at disagreement in the concrete by simply postulating that it has to be overcome; whereas deconstruction never seriously considers that at least some of our differences can be resolved. In no case does anybody seriously look into the sociology and politics of disagreement in the concrete, i.e. into how disagreements get handled in our lives and why they are so treated.

My suggestions as to how this might be accomplished are unabashedly drawn from the all-too-little appreciated work of the American political theorist, William Connolly. In an effort explicitly inspired by Wittgenstein, using a conceptual tool finely honed by W. B. Gallie, the notion of an essentially contested concept, Connolly has shown how it is possible for one to distinguish between legitimate disagreements and those which are the result of mere whimsical contrariness. Legitimate disagreements arise when the parties involved can be shown to be arguing from opposed perspectives on an issue that can be shown (i.e. according to criteria developed by Gallie) to have an ancestry in a richly ramified tradition, which may be said to come into conflict with itself as paradigm cases cease to be shared. What I have done proceeding from Connolly is (1) to show how this notion of an essentially contested concept illuminates the political character of some of the disagreements we find ourselves in and (2) to show how essentially contested concepts must be part of the natural history of a rule-following creature. I have shown, as Connolly has not, just where the Wittgensteinian roots of his position lie and also how that position can cast light upon some of the intellectual contests in which we find ourselves.

Both efforts are in aid of de-mystifying a whole series of notions that mere mortals as well as philosophers and social scientists have long reified: morality, politics, rationality. No small part of my aim has been in aid of showing how they are not detached compartments in our lives but that they pervade it. It is an effort which owes much to the work of Alasdair MacIntyre as well — a debt that I readily acknowledge. The corollary of this view is that there is no "outside" with respect to politics and morality; for in confrontations over these issues my "politics" is often your "morality". This of course runs counter to the ideals of modern social science. Social scientists are always advocates and the sooner all of us realize that the better.

Another important consequence of realizing the role of legitimate disagreement in human life is that we come to be in a position to distinguish between those debates we can hope to resolve, those which we cannot and those in which we can at least establish that both parties are in good faith — which means that they can remain on speaking terms by agreeing to disagree. Agreeing to disagree is in fact recognition that our opponent argues in good faith and that tends to avert us from banal ad hominem vituperation to concern with modes of persuasion. In the concrete I have found that articulating the essentially contested character of, say, the concepts employed in often heated debates with respect to new technologies such as artificial intelligence has proven useful in helping the participants in those debates understand why their opponents are so intractible.

In the essays below, then, I have both applied the Connolly approach in areas where he has not, with a view to bringing out the political character of such things as musical controversies and museums; but I have also shown how Connolly's approach is continuous with Wittgensteinian's later philosophy. I take it that this is important from the point of view of Wittgensteinian studies because, unlike traditional hermeneuticists, some of whom (such as Peter Winch, Tore Nordenstam and Stephen Toulmin) would term themselves Wittgensteinians, my view of the matter entails that hermeneutics (i.e., explorations of how we run afoul of problems of meaning) is not merely a matter of taking a look and articulating what everybody knows anyway, but also of showing us how the tendency to reification (essentialism) is immanent in language itself and thus a permanent temptation.

With respect to the future of philosophy I have suggested that it is immanent in the present but that the dynamics of development are rooted in the past. Several of the essays bellow reflect my particular way of making philosophy into a humanly useful activity by helping scholars in other areas articulate the problems and the methods in their fields — a task that Dewey would have termed "reconstruction" but that today is more likely to be called "hermeneutics". For example, the rambling article which I co-authored with Martin Brody of Wellesley College's Music Department, which serves as a kind of postscript to the volume, is a sampling of the fruits of a seminar on musical culture that we conducted privately between 1980 and 1982. Our aim in the seminar was to explore, not music *and* culture but the ways in which culture finds musical expression and vice versa. The premiss was that intellectual history treats the theme at an all-too-superficial level and music theory hardly treats it at all. Such co-operative ventures using concepts drawn from the arsenal of analytical philosophy, the history of ideas and the philosophy of science to elucidate real problems in other disciplines represents one of the most fruitful forms that current cultural conversations can take. Its aim is to enable all parties to the discussion to articulate aspects of their own discipline which they might not otherwise be able to put their finger upon and to give them the courage to do so. From the point of view of "professional philosophy", the notions which are employed will surely seem naïve but that is just what is the matter with professional philosophy: it has tended to sophisticated technical triviality at the price of making itself incomprehensibly irrelevant to the kind of conceptual problems which actually arise in other disciplines.

This attitude to philosophy was instilled in me by my first teachers at St. Anselm College and later fostered by Stephen Toulmin as well as through many years of association with the Boston University Center for the Philosophy and History of Science. Dominique Lecourt's Centre Cavaillès of the Collège Internationale de Philosophie (formerly the Centre d'histoire et d'epistemologie of the University of Picardie) has supplied another context where I have developed these interests and talents. So it is not accidental that two of the papers collected here were originally presented to Wittgenstein colloquia there in 1983 and 1986. Santiago Ramirez of the Autonomous National University of Mexico, another member of the Centre Cavaillès, was another of my interlocutors in interdisciplinary efforts while I was visiting professor in his Department of Mathematics there.

However, my most important efforts in being philosophically interdisciplinary have been at the Swedish Center for Working Life in Stockholm. It was there in the context of Bo Göranzon's project, "Education, Work, Technology" that I became involved with questions relating to the social consequences of new technology, in particular "expert systems". This project has been raising questions about the nature of skill and experience in work and how the implementation of new technology affects skill and experience in various working life situations. The question which ultimately brought me to Sweden was "what is skill after all?". It is a classical philosophical one inasmuch as it is perfectly simple, yet defies any sort of simple answer. Researchers in the group had all sorts of examples of how skill functions in working life in the form of detailed case studies. What was lacking was the kind of frame of reference with which they could present this material convincingly. By the time I arrived they had already hit upon Polanyi and Wittgenstein and were wrestling with them in their efforts to construct an epistemology of work (a task especially dear to two of my first philosophy professors). What was required was a philosophical orientation of the sort that only a trained philosopher could deliver but it had to be one that was open to the kind of task they had before them. They did not need to learn what to think but of how to deepen their insights and present them in a constructive way. Thus, my work at the Center, which is briefly sketched in the paper on "Tacit Knowledge" below, is typical of the hermeneutic and heuristic role that I continue to play there with more success than I myself would have expected before I arrived in Stockholm. It is an indication of one (I emphasize) sort of cultural conversation that philosophers can involve themselves in. Since I share Evelyn Fox Keller's belief that the only realistic way into the future is on the basis of encouraging what is healthiest in our current practices, I offer the results of these experiences for consideration to the philosophical community at large.

In the Toscana ALLAN JANIK
March, 1988

CHAPTER I

STYLE AND IDEA IN THE LATER HEIDEGGER: RHETORIC, POLITICS AND PHILOSOPHY

After fifteen years Robert Minder's learned analysis of Heidegger's lecture on Johann Peter Hebel remains by far the most persuasive effort to link Heidegger's thought to Nazism.[1] Minder does this through a deft exegesis of the style of Heidegger's essay, which leaves no room for doubting that there are highly significant connections between the later Heidegger's style and a certain conservative, Catholic-peasant ideology which could and in fact did easily become Nazified. Proceeding from the (dubious) premiss that Adorno's philosophical analysis of the "Jargon of Authenticity" accurately exposes the *philosophical* shortcomings of Heidegger's *Seinsdenken*, Minder considers the later Heidegger's *literary* affiliations with the same end in view. Heidegger's style goes through four phases. First, we have his early student years when his writing is aridly academic. In his second period he writes *Sein und Zeit* in a style that adapts Expressionism to philosophical prose. The third phase coincides with his overt support of the Hitler regime as rector of the University of Freiburg in 1933 and is, consequently, overtly Nazi. The final phase, Minder argues, represents a muted depoliticization of Nazi pseudo-Romantic obscurantism. This represents an extension of the thesis first advanced in Paul Hühnerfeld's trenchant study *In Sachen Heidegger*.[2] On the basis of a wide variety of stylistic and substantive parallels with Nazi and proto-Nazi authors Minder concludes that Heidegger's style itself provides us with yet more reason to reject Heidegger's philosophizing. This is surely a very tempting reading of Heidegger's late prose. Passages like the following, which is the denoument to the Hebel lecture, seem to invite such a reading:

> Der Mensch spricht aus jener Sprache hinaus, der sein Wesen zugesprochen ist. Wir nennen diese Sprache die Muttersprache. Eigentlich spricht die Sprache nicht der Mensch. Der Mensch spricht erst insofern er jeweils der Sprache ent-spricht.[3]

Heidegger does indeed seem to reify the notion of language and, what is worse, to shroud it with mystery in the manner of just those Romantic authors who were most appealing to the Nazi ideologists.[4]

Moreover, his reference to dialect (*Muttersprache*) as the origin (earlier in the lecture as the *Quelle*), as Minder emphasizes, is typical of *völkisch* rhetoric. The same is true of the peculiarly repetitive usage of Sprache and its cognates. So there is no question whatsoever of the parallels that exist between *völkisch* writers and the later Heidegger both in style and tone. However, I suggest that a yet closer view of Heidegger's style, one which links it to his overall philosophical positions and strategies, will yield a different perspective on the matter. With this in mind I propose to undertake a critique of Minder's approach to Heideggerian philosophy through Heidegger's style. Precisely because Minder's is a highly informed perspective on Heidegger, a critique of his view of the later Heidegger will help to illuminate some of the enormous difficulties surrounding the evaluation of the most influential German thinker of our time. In order to persue this critique it will be necessary to begin by enumerating Minder's chief claims.

The first aspect of Heidegger's Hebel lecture which disturbs Minder is that it does not accurately represent the historical Hebel.[5] Basically, Heidegger fails entirely to represent Hebel as a champion of Enlightenment cosmopolitanism despite his commitment to *Alemanisch* dialect. Minder finds it difficult to come to any other conclusion than that Heidegger is completely lacking in respect for historical accuracy in the light of his cavalier treatment of Hebel and that this disregard for fact exposes Heidegger as a mere sophist. He is less the true champion of the splendor of the simple peasant life than he is the pedantic peasant schoolmaster become mystagogue. The second part of Minder's claim is that Heidegger's disregard for fidelity to history is rooted in his commitment to linguistic nationalism with all of the spurious metaphysical baggage which has accompanied that concept in the last two hundred years of German intellectual history. Minder's wrath falls upon Heidegger for erroneously associating Hebel with this family of ideas. Indeed, much of Minder's indignation arises because Heidegger misrepresents factual issues such as the scope of Hebel's reputation — he was known throughout Europe — and his strongly pro-Enlightenment cosmopolitan values. Minder sees Heidegger as extolling Hebel as a champion of the *Alemanisch* dialect, which Hebel certainly was, for reasons which are totally foreign to Hebel's world view. Heidegger's interest in Hebel follows from his concern with maintaining the purity of the German language in the face of foreign contamination, i.e. the last stage of an effort which began with Luther to keep the German language pure of malaign foreign influences. A third part of Minder's argument claims that Heidegger's approach to Hebel is typical of the *völkisch* thinkers of the thirties both in its "Wagnerized" alliterative rhetoric and in its exultation of organic social relations, *Gemeinschaft* and the like. Heidegger's emphasis on dialect as "Muttersprache" and "Quelle" in the Hebel lecture unmistakably links him to the *Alemanisch Blut und Boden* proto-Fascist and Nazi writers of the period, from whose writings Minder musters an impressive array of convincing parallels. For example, he cites a passage from one Wilhelm Schäfer which certainly shares some family resemblances with, say, Heidegger's *Der Feldweg.* Schäfer writes,

> die muntere Wiese dem strengen Schwarzwald entspringt: Da gingen dem Knaben die Wege in fröhlicher Freiheit, da waren die Wolkenweiten über die grünen Gebreite bis hinter die blauen Fernen gebogen. Da sangen die Vögel zur Arbeit, da war ein emsiges Landvolk im Kreislauf der Jahre geborgen.[6]

In his little philosophical fable Heidegger writes,

> Indessen begannen Härte und Geruch des Eichen holzes vernehmlicher, von der Langsamkeit und Stete zu sprechen, mit denen der Baum wächst. Die Eiche selber sprach, dass in solchem Wachstum allein gegründet wird, was dauert und fruchtet: das Wachstum heisst: der Weite des Himmels sich öffnen und zugleich in das Dunkel der Erde würzeln; das alles Gediegene nur gedieht, wenn der Mensch gleich recht beides ist: bereit dem Anspruch des höchsten Himmels und aufgehoben im Schutz der tragenden Erde.[7]

So, there can be little doubt that there are similarities in style and tone between their works. Further, when Heidegger identifies his village, Messkirch, as the "Ort des Ursprungs" (literally, the place of origin) or the "Gegnet des Wortes" (literally, the territory of the word, employing the local variant *Gegnet* for *Gegend*) Minder claims Heidegger is doing little more than repeating Nazi clichés extolling the rural at the expense of the urban and ultimately the visceral at the expense of the rational. Minder's fourth claim, then, is that Heidegger has so sanctified a 'poetic' conception of language that both rational criticism and literary artistry as they are normally understood are no longer possible. The essence of language has become *litany.* As we have seen in the passage cited above from the Hebel lecture there are certainly prima facie grounds for such an assertion. Minder's final claim is that this represents the ultimate betrayal of Western values, for it amounts to nothing less than moral indifference. Thus, Minder's literary analysis of Heidegger's Hebel lecture becomes the basis for condemning Heidegger's philosophy as a whole. It is this latter move that I wish to contest.

There can be no doubt whatsoever that we know a great deal more about the background to Heidegger's style after Minder's analysis than before and in two senses: first, we are simply in possession of a great deal of data that was not previously available; secondly, he has raised nearly all the crucial questions which must be answered if we are ever to appraise Heidegger's thought accurately. However, this is not to say that we need accept his views on the subjects he has discussed. Before proceeding to look at the problems with Minder's position and suggesting alternatives it is crucial to have a sense of the strengths of his analysis, which are at one and the same time weaknesses in Heidegger — weaknesses which Heideggerians have been altogether too hasty in passing over — and, thus, points of departure for a proper evaluation his thought.

First off, the Hebel lecture does indeed look like a piece of hack literary exegesis as do many of both Heidegger's poetic and philosophical 'interpretations'. We need only call to mind his readings of, say, Anaximander or Georg Trakl for examples of Heideggerian exegesis which are highly contestable.[8] If this is all that they are, it is difficult to avoid Minder's conclusions. Nevertheless, if these works are more than this, Heidegger has given us precious little inkling of just what he has been up to. Secondly, Heidegger's approach to Hebel does indeed seem Romantic. There can be no doubt that Heidegger is neither a champion of the Enlightenment nor of liberalism — nor does he go out of his way to identify Hebel with these movements. Further, we cannot deny that Heidegger's works surround themselves with an aura of authoritarianism. Thirdly, the stylistic parallels which Minder has pointed out between Heidegger and Anton Gabele as well as other Nazi and proto-Nazi writers of the twenties and thirties leave no question whatsoever in their wake that the later Heidegger's characteristic modes of expression are clearly related to, if not derived from, theirs. Both the organic metaphors and the pathos of these *völkisch* thinkers is parallelled in his late works. Fourthly, Heidegger does indeed seem highly concerned about the way in which we have come to forget the original meanings of words such that he can be plausibly represented as preoccupied chiefly with insuring the purity of the language. Further, to anyone trained in historical linguistics his etymologies seem to be little more than pompously glorified folk etymology.[9] Fifthly, it is distressing in the extreme that Heidegger not only failed to make crystal clear just what the method was that informs his later work. He also never did so much as to reject Romantic, 'metaphysical' and even Nazi readings of his works. All of the accounts we have from him of his own activities, including the celebrated *Spiegel* interview, are less than illuminating to the skeptic, for Heidegger never attempted to explain or excuse the support he extended to the fledgling Hitler regime between 1933 and 1935.[10] Sixthly, the rejection that the later works can be characterized as true or false (not explicitly discussed in Minder's essay but nonetheless implicit in it) invites the sort of sociological explaining away of Heidegger's views that we find in Minder's analysis:[11] Minder tries to make what he takes to be patent falsities and absurdities in Heidegger's thought comprehensible by asserting that Heidegger was merely reflecting the attitudes of his social group (the peasantry), his professional peers (university professors) and his generation (the *Frontkämpfer* of 1914). To put the point somewhat differently, because Heidegger put himself beyond truth and falsity, he both talks nonsense and is, therefore, not to be taken as he would have himself taken, i.e. as a serious philosopher. Finally, it does seem as Minder alleges that poetry has become a secular religion for the later Heidegger; his later prose does indeed appear to be little more than litany and little more than pure verbosity. All of these considerations suggest that Minder has to be taken very seriously indeed, for he makes a strong case for each of these allegations. However, to have made a strong case is not yet to have made a *compelling* one. These are the right questions to ask about Heidegger but they can be answered differently from the way Minder has replied to them. This is what I propose to do in what follows. It is necessary to assert at the outset that all of these issues are remarkably difficult to resolve in that absence of a definitive biography of the philosopher, for we must bear in mind that we do not as yet, pace Minder, even have a solid grasp on the precise nature of Heidegger's relationship to Nazism, as George Steiner has emphasized.[12] The great difficulty in stating Heidegger's views without relying upon his own formulations is yet another difficulty that cannot be passed over in silence. Since it is central to Minder's case that Heidegger's later thought is essentially a muted depoliticized version of an earlier commitment, what we discover in the definitive story of that commitment will have bearing on our evaluation of his later philosophy.

What, then, are the problems with Minder's claims? What might lead us to question their soundness? If Minder's account is not wholly accurate, how ought we to understand Heidegger's style? To answer these questions we shall have to call the assumptions which underlie Minder's perspective on Heidegger into question. The most effective way of accomplishing this will be to provide alternative readings, albeit speculative ones, which will have the force of underscoring the speculative elements in his account of the later Heidegger.

We can begin by asking just what Heidegger's suspicious silence indicates concerning his Nazi past. Many critics have simply assumed that his silence was due to the fact that he could neither explain nor justify his actions in the mid-thirties. Doubtless, this is the easiest way to understand the matter. However, this does not follow from the mere fact that he never made an effort to clarify the situation. There is no question that they require explanation but from the perspective of a conservative old German professor — one, we should recall, who was not allowed to teach for some years on account of his actions in the mid-thirties.[13] — with a highly developed sense of decorum, it *might* seem best not to dignify what he perceived as slanderous attacks on his integrity and let histroy speak for itself. Was this the case? It *may* have been. Unfortunately, the record is remarkably unclear about the significance of his actions. What is absolutely certain is that his literary executors do no one a service by restricting access to his papers where we might find some illumination about his intentions, commitments and actions. In any case, it is important to remind ourselves continually how much remains to be learned from a definitive biography of the philosopher, one which, unfortunately, does not seem to be forthcoming.

The next question we can raise is whether the Hebel lecture was ever intended to be historically informative. There is no question whatsoever that it looks as though it were. Yet, it does not follow that Heidegger is only talking about the historical figure just because he seems to be doing that i.e. he might be doing lots more. The whole point of the lecture is to establish that language does more than communicate information.[14] I am suggesting that the lecture may be a way of doing something with words that is other than conveying information. Indeed, Minder himself comes close to what I am suggesting without knowing it when he describes Heidegger's lecture as a work of *edification.* I shall argue below that there is good reason historically and philosophically for considering Heidegger's later writings as a peculiar development of traditional spiritual techniques. In any case, the point of edifying rhetoric is to put the listener in a position to see something that he did not see previously in such a way that the acquired insight will have an *impact* upon his beliefs and actions. Perhaps the simplest way of putting this point is as follows: the literal truth of the sentences is not the primary consideration in a sermon, for those sentences are not propositions. This may seem on the face of it philosophically offensive but the notion that philosophy is principally concerned with sentences which are neither true nor false makes perfectly good philosophical sense from the perspectives of thinkers as different as Wittgenstein and Collingwood, thinkers not normally associated with Heidegger or Nazism. For Wittgenstein everything that can be said clearly can be expressed in a true or a false proposition; however, no proposition can say that, for it can only be shown. The task of philosophy is to make clear the distinction between what can be said and what can be shown in statements which are neither true nor false but which stimulate self-reflection such as aphoriams or questions.[15] In a similar vein R. G. Collingwood insists that truth and falsity are the properties of scientific propositions which themselves rest upon assumptions which are not propositions and therefore neither true nor false.[16] Thus, for both Wittgenstein and Collingwood, like Heidegger, philosophy is essentially a laying bare of the presuppositions which make truth possible rather than a set of true propositions. If Heidegger is in fact doing something like delivering a sermon we should not expect truth and falsity of his sentences to be central concerns. Hence, the first set of objections Minder poses to his lecture becomes less significant (note I do not say insignificant) than it might be. This leads us to an alternative reading of Minder's Romantic interpretation of Heidegger.

There are indeed a number of organic metaphors in the Hebel lecture and throughout the corpus of the later Heidegger but the existence of these metaphors alone is insufficient evidence for identifying Heidegger with the kind of commitments characteristic of Nazi and proto-Fascist organic theorists. If nothing else is clear in *Sein und Zeit*, it ought to be obvious that he is trying to purge metaphysics of precisely the sort of reifications upon which organic theories of society rest.[17] Organic theories attempt to explain the way we find the world in terms of an immanent teleology within reality; reality is a process striving to accomplish something or realize an idea or pattern. This alone is ultimately real. However, for Heidiegger to assert this is to make the fundamental error of confusing Being with a type of entity; it is to think of what is ultimate as the precondition in terms of which things are possible as opposed to that precondition which makes action possible (see the discussion of the distinction between *Vorhandene* and *Zuhandene* below). The obvious objection arises here that Heidegger simply should not have employed such metaphors as "Sprachgeist" if he was not invoking the organic theory of society. Two considerations are relevant to this objection. First, there is historical evidence that Heidegger was concerned with poets and thinkers in the *Brenner* Circle, whom I shall discuss below, who used similar expressions, especially landscape, independently and in explicit criticism of *völkisch* ideology, and indeed in opposition to it. Secondly, if Heidegger is in fact sermonizing in some sense or other, his goal will be facilitated by expressing himself in language familiar to his audience. There can be little doubt that metaphors for depth such as 'Quelle' and 'Ursprung' would be familiar to his listeners. This leads directly to the question of whether the stylistic parallels with *völkisch* writers implicates Heidegger as a *völkisch* thinker.

It is of course possible that Heidegger's style was derived from the likes of Anton Gabele as Minder claims.[18] If we knew that for certain the literary question would be closed. In the absence of established fact these stylistic similarities remain circumstantial evidence. Once more an objection might be raised to the effect that, given the plethora of evidence Minder musters, it seems difficult to conceive an alternative to Minder's view of the matter. It is indeed difficult but it is by no means impossible. The crucial clue lies in the edificatory character of much of the later Heidegger's writing.

Nearly all of the stylistic devices that Minder associates with Wagner and the *völkisch* thinkers have long been stock-in-trade to preachers, retreat masters and novice masters. Assonances, folk etymologies, litanies, forced symmetries of linguistic form and the like have been the mainstays of a certain type of preaching at least since the Counter-Reformation. Is it altogether accidental that the figure who brought this technique to the level of an art form, Ulrich Megerle, who gained fame as the Habsburg Court Preacher in Vienna from 1677—1709 under the name Abraham a Santa Clara, was also a son of the very same Messkirch where Heidegger was born and indeed the subject of one of Heidegger's late essays? The answer to that historical question is unclear but it is certain that all of the rhetorical gimmicks we find in

Wagner and the *völkisch* writers Minder discusses were well known to Abraham, whom Hugo Hantsch has described as the greatest master of German prose between Luther and Goethe;[19] so they did not have to be invented by Wagner and Co. We find, for example, that Abraham uses puns and curious etymologies in his work in a way that is, at least stylistically, similar to Heidegger:

Wie aber soll ich die Welt nennen? Holla Welt! ich frag dich? Was vor einen Titl soll ich dir zueignen? Wer bist du? Sags her hast du es verstanden? Holla! Die Welt antwortet mir durch den Echo in dem Wörtlein Holla! *Olla*, das heisst auf Lateinisch ein Hafen oder ein Topff so ist dann die Welt ein irdischer Topff" Ja ja in diesen Topff ist ein wunderlich *Allapatrida*.[20]

This last word is a pun on the Spanish word for mish-mash. Abraham was no less fond of alliterations than Wagner: Frei, frisch, frech, fröhlich und freundlich ist die Jugend, wesenthalben Jugend und Tugend selten beisammen.[21] This, then, helps to explain how Heidegger's works could be stylistically similar to Wagner's as Minder insists; while Heidegger can unequivocally condemn Wagner as a thinker (see below), a fact of which Minder does not appear to be aware. To appreciate the reasons why Heidegger would want to exploit this literature will require both a discussion of his philosophical strategy and why what are traditionally known as 'spiritual exercises' are relevant to it. But that is to run ahead. The point is that the stylistic similarities that we find between Heidegger and the *völkisch* writers with whom Minder associates him can be explained by reference to a common heritage as well as by reference to an influence the *völkisch* thinkers exerted upon him.[22]

The next issue we must address is Minder's claim that Heidegger's concern for linguistic "origins" is part of his battle against the corruption of the German language through the introduction of foreign words into the German vocabulary. In fact Heidegger never makes this sort of complaint. What he does complain about is the way in which philosophical words have been translated from ancient Greek into Latin and then into the vernacular, for in the process their original connotations have been lost. Thus, for example, *zōon echon logon* becomes *animal rationale* with the result that thought rather than speech tends to become the defining characteristic. Be that as it may, the concept of purifying language from its foreign accretions is one which is wholly foreign to his philosophical program. He makes it abundantly clear in such writings as *Nietzsche*, "Die Zeit des Weltbildes" and *Gelassenheit* that there is a certain inexorabe character to history, which for Heidegger is the history of Being rather than empirical history,[23] that makes genuine historical developments unavoidable. Like Hegel, Heidegger is firmly committed to the idea that philosophy cannot transcend the age and, like Wittgenstein, he is committed to the notion that philosophy neither can nor should aspire to change the world. The kind of *restitutio in pristinum* which has been the goal of Luther et al. is something that Heidegger considers to be impossible.

Finally, it never occurs to Minder to ask whether there might be any foundation to Heidegger's poetical notion of philosophy. He never asks whether Heidegger might have reasons for making poetry into something sacred or for celebrating this in litany form. This is principally because Minder makes no effort to come to grips with the kinds of things Heidegger claims to be doing philosophically in his later works. Because Minder sunders the style from the substance of what it is tailored to convey, he distorts both. It may seem strange for a philosopher to make so much of poetry but it is worth recalling Wittgenstein's dictum "Philosophie dürfte man eigentlich nur dichten".[24] Doubtless, Wittgenstein did not intend to defend a Heideggerian concept of philosophy when he penned that statement, but he did mean it seriously. However tortuous the bends and twists on Heidegger's path of thinking we can only pass judgment upon it once we have traversed it. To do this we have to be prepared to meet Heidegger on his own ground, something Minder has not done. To have said this is by no means to have asserted that there is anything simple about finding, let alone following or assessing, the path of thinking, as Heidegger terms his later philosophy. In order to do that from the points which Minder wishes to address we must begin by looking into the philosophical and historical reasons which might make the language of edification appropriate and plausible for the message which Heidegger wishes to convey. To do full justice to this task would take a full length study; however, it is possible in the scope of a single essay to identify the sorts of issues which are germane to that study. These are above all issues about the relationship between his style and the ideas he wishes to express through it. Before proceeding to clarify that relationship I must make the sorts of things I assume to be the case about Heidegger's thought absolutely clear.

The assumptions which determine my approach to Heidegger are the following:

(1) there is only one Heidegger; *Sein und Zeit* and the later writings do not represent different positions but different avenues of access to what is fundamentally the same position;
(2) we need not always be committed to Heidegger's mode of expression; sometimes it is very important to recast his thought in different language even though this will not do in all cases (it goes without saying that this must be done cautiously);
(3) it is possible to understand Heidegger without agreeing with him; his position on, say, the end of philosophy is intelligible apart from any leap of faith;
(4) it is not clear that his thought is all of one piece as he represents it — one need not accept or reject it as a whole, i.e. elements of his position can be assessed separately from each other;
(5) his style is *deliberately* elusive and complex, so translations only approximate what he says and often obscure his puns etc.; it is usually more profitable to discuss his German texts than to try to translate them. Thus, I have left most of the texts in German.

The question I wish to proceed from will allow us to focus sharply on the relationship between philosophical substance and literary expression in Heidegger's work: why is it a mistake to read Heidegger as a Romantic or as an Expressionist? Romanticism in the context of German intellectual history and especially as regards language, as Minder himself rightly points out, involves a commitment to the notion that there are spiritual entities such as the *Volksgeist.* As spiritual entities these objects are unobservable; they provide the explanations and meaning of what is obervable. In short, they are variations on Platonic Ideas.[25] However, the whole force of Heidegger's distinction between *Vorhandende*, what is present to hand or present in the manner of an object of observation, and *Zuhandene*, what is ready to hand or available for use as a tool is to undermine such Platonizing.[26] The critique of traditional epistemology in *Sein und Zeit* is an attack on the Platonic and Cartesian construal of the object of thought as something utterly other than the sort of things we encounter in everyday life as tools or obstacles. First we conceive of entities as objects of thought, then we enquire into the origins, foundations, ultimate causes or whatever of both those entities and the "I" that knows them, which by parity of reasoning must be an entity of the same order as the objects it knows. Thus, a chain of reasoning is set in motion which leads us to think of reality as something other than the workaday world of tables and chairs, hopes and fears, lovers and enemies, as constituted by ahistorical things which exist over against a knowing subject rather than human activities which always take place within a context and are thus radically historical. This is the force of Heidegger's 'ontological difference'.[27] Heidegger wants to distinguish Being from beings or entities because he thinks that a fatal confusion slips into metaphysics when we take Being to be the source of the possibility of the existence of 'things', which are 'out there' in the world rather than that which is the absolute presupposition of the kind of entity which *acts*. It is crucial for philosophy because philosophers have traditionally aimed at a complete and consistent causal account of reality in terms of objects present to hand in such a way as would radically transform the nature of the human world into something in which the peculiarly human element, intentionality or being in a situation, is entirely lacking. Thus, the world of being-in-a-situation is transformed into an epiphenomenon precisely by requiring that it be 'explained' in terms of eternal immutable objects or laws in such a manner as to conceal the fact that intentional activity is the primary datum in the human sphere. This is the force of the distinction between *Vorhandene* and *Zuhandene* and the metaphysical explanation of our tendency to reify. To the extent that we make Platonic ideas into what is 'really real' we transform words into things. The whole force of *Sein und Zeit* is to combat that reification, to stand the distinction between things and actions on its head and, thus, to rehabilitate the world in which we really exist as agents as ultimate reality. This distinction implies that philosophy as a theory of ultimate reality or as 'foundationalism' is a conceptual impossibility, for activity is prior to theorizing and incapable of being comprehended by it. To the extent that the Romantic theory of knowledge rests upon precisely that kind of reified theoretical language the whole thrust of Heidegger's early work is to reject it unqualifiedly. Thus, Heidegger's use of words like "Quelle" and locutions like "die sprache spricht" or the more notorious "das Nichts überhaupt nichtet" cannot possibly have the sort of connotation that Carnap[28] and Minder have assigned to them. *Das Nichts* is not to be understood as a thing. Indeed, one reason for using such a locution is to say that nothing is not an object like the square root of two or a cat named Jezabel or Albert Einstein.[29] To put the matter another way, the whole point of using such an odd expression is to establish that *das Nichts* is not a referring expression. This will perhaps become clearer as we examine the gap that seperates *Sein und Zeit* from Expressionism.

There can be little doubt that there are numerous surface similarities between the style of *Sein und Zeit* and typical Expressionist rhetoric.[30] The contrasts between the authentic (*eigentlich*) and inauthentic (*uneigentlich*), the notion of being-there (*Dasein*), the notion of dread (*Angst*), death (*Tod*) and resolve (*Entscheidung*) and the notion of thrownness (*Geworfenheit*) all smart of the Expressionist stage of the early twenties, as do both Heidegger's numerous neologisms such as *zuhanden, Dasein* etc. and redundancies like *die Sprache spricht*. The emphasis on *Angst* and the notion of death as it occurs in the expression being towards death (*Sein zum Tode*) especially seems to be indicative of a link to that movement whose most representative symbol was the Cry.[31] Walter Kaufmann has pointed out that Buber's penchant for unconventional expression in *Ich und Du* formed both the first link between the rhetoric of Expressionism and philosophical style and paved the way for the avant garde's enthusiatic reception of *Sein und Zeit*.[32] It is important because it helps us to grasp why Heidegger absolutely and unequivocally rejected the Expressionist reading of his book from the outset, for after the publication of *Ich und Du* there was a predilection in the German public for reading what was new and innovative in precisely that way.[33] Indeed, he anticipated objections to the inherent obscurity of his style by insisting that this was endemic to the sort of undertaking with which he was engaged, citing the stylistic obscurities of Plato's *Parmenides* and the Fourth Chapter of the Seventh Book of Aristotle's *Metaphysics* as cases where earlier researchers into the problem of Being were led to "unlovely" expression[34] by lack of an appropriate vocabulary and grammar. He rightly suggests that a comparison of the aforementioned metaphysicians' prose with that of Thucydides will be sufficient to establish his claim. More important philosophically is his distinction between two modes of phenomenological description, the *existential* and the *existentiel*. The latter is what has come to be most closely associated with Existentialism and bears directly upon Expressionism which drew attention to the same features of human experience as the so-called Existentialists, i.e. to the role of pain and suffering in defining what it is to be human. Thus, the Existentialist writers would discuss human beings as defined by their actions, commitment and feelings rather than on the basis of some theory of human nature. Heidegger is not opposed to this but it is not his main concern. His main concern is, given that the kind of picture that, say, Kierkegaard draws of what it is to experience the world humanly, to ask how is this at all possible? An *existential* phenomenology or a fundamental ontology basically proceeds from the question what must Being be like if the being for whom Being is always a problem is the way it is? This is clearly not what Expressionism is about despite any surface similarities. Heidegger is always a metaphysician. Further, his distance from Expressionism becomes apparent when we examine his views about language, for we discover already in *Sein und Zeit* that he so conceives language that his message requires a most peculiar medium.

His mentor Edmund Husserl had established that an adequate account of meaning must disclose the significance of expressions, not in relation to some theoretical criterion of demarcation but to the world of lived experience (*Lebenswelt*).[35] Heidegger's task was nothing less than an exploration of the absolute presuppositions of lived experience, i.e. the Being of the life world. Husserl had sought to establish that there was no such thing as pure consciousness, that consciousness was always consciousness *of* something. Heidegger, while rejecting the notion that there is such a *thing* as consciousness (the word *Bewusstsein* is entirely foreign to his work — in this he is closer to William James's radical empiricism than his is to the later Husserl with his transcendental ego), went yet a step farther in insisting that "Reden ist Rede über".[36] Not simply language but spoken language is locus of intentionality and, thus, of metaphysical significance for Heidegger. Here we arrive at the kernel of his problem. If discourse is always discoursing upon some topic or other, discoursing about language will have a curious self-referentiality. As such it takes us farther away from things themselves rather than bringing us closer to them. Heidegger's task is not only to render this circularity less than vacuous by showing us what language is in *action* rather than theorizing about it. This is nothing more than to transform that circularity into something positively illuminating — something which ordinary factual discourse as understood, say, by the positivist or Neo-Kantian philosophers of the day could hardly accomplish. But this is not to say that there are no language games available for that purpose. His task is at least in part to create a language that can *do* something, to wit: assist us with the act of self-reflection which will permit us to 'see' ouselves as makers of speech acts rather than produce a theory of speech acts. In Wittgenstein's parlance, it is far more a matter of showing what it is that language does than saying something about its 'nature'. More to the point — and this is one of a number of issues upon which Heidegger is almost entirely at one with Wittgenstein[37]

— discoursing *about* language invites us to think of language as another object and thereby distorts the way it comes to *constitute* the world in which we find ourselves. In Kantian terms language is transcendental. From the very first moment of our existence we are 'thrown' (*geworfen*) into a network of meanings.[38] Our very reactions to our environment — fear, joy, anger and the like — are only possible because we have 'internalized' these networks in the process of learning to get around in the world such that we are able to recognize situations as dangerous, delightful, insulting etc. The same is true of our ability to plan future courses of action. Even our ability to maintain silence is rooted in discourse for Heidegger[39] — the important difference between them is that for Kierkegaard this has moral (*existentiel*) significance; whereas for Heidegger, who always asks how is the *existentiel* at all possible? How is it of metaphysical (*existential*) significance? Metalanguage, far from drawing us closer to an understanding of language, focuses our attention from the way in which we are enmeshed in language. Indeed, Heidegger emphasizes that language is to be grasped in the concrete rather than in the abstract (i.e. theoretically) when he insists that it is in discourse, in 'saying' and 'speaking' rather than through an analysis of the nature of propositions à la Russell and the the early Wittgenstein that we have to understand language. *Dasein* is constituted through language, which is nothing other than to say that its primordial mode of being — action — requires language. In the lived-world language and action are inextricably united conceptually (equiprimordial). They are unintelligible taken apart from one another. Thus, we are unable to conceive of one as prior to the other. Speaking and doing are two sides of the same coin — a fact that the 'propositional' account of language wholly conceals precisely because it proceeds from the abstract understanding of language that is the written word rather than the concrete spoken word. Heidegger's task, like that of the later Wittgenstein, is nothing less than the rehabilitation of the philosophical significance of the spoken word. His problem is, at least in part, that of conveying an insight into the workings of language in everyday life for metaphysics, in a language which retains the richness and vitality of pre-theoretical speech. He has to teach us to take a second look at the conceptual bond between acting and speaking, which we take for granted in our everyday affairs and which is prior to theorizing. Any theory is built upon assertions or propositions which Heidegger has shown to be derivative from the world of equipment (action).[40] Theories have the effect of transforming something which is ready to hand into something present to hand. However, that speech which makes the ready to hand available for discussion in the first place is plastered over, as it were, in the transformation. So, all theory conceals its foundations in human activity but theories of language do so in a double sense. To put the matter another way for there to be theories at all there must be shared meanings. So a theory of meaning presupposes that there is sufficient understanding available to articulate a theory. To theorize about meaning, then, is in some way or other to assume what we wish to demonstrate. This is why Heidegger ignores theory for the sake of re-directing our attention. In short, what Minder and others take to be the pathos implicit in the language of *Sein und Zeit*, I take to be part and parcel of an effort to redirect our attention to precisely those elements about language which most impressed the author of the *Philosophical Investigations* and which presented Wittgenstein with just as many problems with regard to expression.[41] So, there are good reasons why the style of *Sein und Zeit* ought to be out of the ordinary. But this is not yet to explain the stylistic idiosyncrasies of the *later* Heidegger.

At this point we must ask why it is that Heidegger would want to exploit the repertoire of edification associated with the notion of spiritual exercises. In order to answer this question we shall have to explore Heidegger's notion of the present situation in the history of Being, on the one hand, and the nature and function of spiritual exercises, on the other.

First off, he clearly sees himself, like Plato and Aristotle, at a turning point in the history of philosophy, one which is monumental enough to count as a new beginning of the subject. The crucial characteristic of our time according to Heidegger is a phenomenon that I want to call paradigmlessness. I have chosen this term to describe what Heidegger designates as the social correlative of what he terms the end of Platonism or the end of philosophy after the usage in Thomas Kuhn's well-known book[42] because, like a Kuhnian paradigm-shift, the transformation of the concept of rationality in the transition from Platonism to post-Platonism involves a *radically* different way of understanding the way things are. We have according to Heidegger two incompatible and incommensurable frameworks not simply for investigating a given phenomenon such as electricity but for grasping reality. Platonism (which must not be confused with the philosophical postion of Plato himself) or metaphysics, is a paradigm of rationality which determines both the practical and theoretical horizons of an epoch in the history of Being (*Sein*). It is no less determining of the nature of philosophical activity and its place in human affairs. This is what has radically changed in Heidegger's view. We will understand the 'paradigm-shift' or what I called the shift to paradigmlessness — that has taken place with the end of Platonism better if we examine the stages in the development of Platonism, a notion which Heidiegger in fact borrows from Nietzsche.

For Heidegger Platonism is the view that there are two worlds, the imperfect world of appearance which we perceive via our senses and the perfect world of reality which we know intellectually and which is the foundation and explanation of the former. In the first stage, which is found in Plato himself,we attain the world of the really real by living wisely, which is possible by examining our beliefs for consistency. The second stage is ushered in by Christiantiy, which is 'Platonism' proper and the means of transmitting metaphysics to the masses, involves understanding that the real world is not attainable now but is promised to those who live wisely and well. Henceforth, Christian theology, whose credentials are determined by its logical coherence, becomes the paradigm theory for rationality. The third stage is Kantianism for which the true world is unattainable and indemonstrable; it is not something we can hope for but rather a moral imperative or duty. Here the substance of the Christian paradigm theory, rational proof of the the existence of God, the created nature of the world, etc., is denied but its function as providing the logical 'foundation' of knowledge and action is retained. The fourth stage is German Idealism, the critique of Kantianism, to which the concept of the true world simply ceases to have a literal meaning, by which I take Heidegger to mean that a secular, metaphorical meaning is sought for theological concepts as in Hegel and Schliermacher. In the fifth stage positivism rejects the concept of the true world as useless and leaves us with appearances alone. The chief claim of the paradigm theory is utterly demolished but the conceptual framework it provided remains. Thus, nineteenth century positivists, having stood the two world theory on its head, will speak confusedly of appearance as the only reality. The final stage is the transition from positivism to perpsectivism in which the last vestige of Platonism, the notion of a world of appearance, is abolished. It occurs in the late works of Nietzsche, in which the notion of ultimate reality and foundations for knowledge are completely rejected. Thus, the last vestige of the framework in which Western thought developed is dissolved. Particular enterprizes, scientific and social, can be carried on from within a particular perspective but there is no overarching order of things or totality. There is no security whatsoever in knowledge or life. Because there is no order at all we are left to our own devices to learn to see as best as we can and to create the best values we can in the absence of any legitimating authority whatsoever. Thus, in Kuhn's terms our situation becomes analogous to that of someone working in physical optics before Newton: "Being able to take no common body of belief for granted, each writer on physical optics felt forced to build his field anew from its foundations".[44] According to Heidegger this is precisely the state of all of our affairs at present. Language, as we shall see, becomes crucial for Heidegger as our only resource in the face of this paradigmlessness.

There are essentially two responses to our situation of paradigmlessness. One is the post-philosophical path of thinking' (*Gelassenheit*) which stoically seeks to comprehend the nature of the crisis we find ourselves in and accepts the challenge to live in a vertiginous foundationless world, a world 'without a why'. The other is manifest in an incapacity to face the void of paradigmlessness, life 'without a why'; it entails a flight into the security of one or another world view. The latter is an homogenization of politics, science, art and social beliefs which plays the role of surrogate in the absence of legitmate authority.[45] World-view is a means to power, an effort to make the world what it is not, i.e. neatly intelligible as a totality. Heidegger finds the metaphysical origin of this notion in the conception of the world as something pictured (*Vorstellung, Bild*). The Era of the World Picture arises at the beginning of the modern period when Descartes conceives reality as an object for a subject. On Heidegger's view this is an implicit challenge to the subject to gain control over the object. Knowing then becomes a mode of attacking Nature for the sake of dominance. Dominance becomes an end in itself. Technology inevitably comes to take the place of wisdom. Traditional metaphysics demands a total explanation of reality but this is no longer forthcoming. All we are left with is a plurality of intellectual and social perspectives, which cannot be reconciled in terms of some 'higher' synthesis. Insecurity in the face of this situation leads those who have not had the courage to face life in the absence of foundations for thought or action to 'will' those foundations where they do not exist. Thus, the polar opposite of Heideggerian

Gelassenheit is nothing other than fanatical rejection of the way things are. I think there are good reasons for believing that the Nazi horrors and his own metaphysical path of thinking are thus intimately related, i.e. as opposite reactions to the way things are in the twentieth century. Moreover, it also helps to explain how it might be possible for someone with Heidegger's background and concerns to confuse his undertaking with Nazi claims to revitalize the socio-political order, for each develops an aspect of a remote rather than a proximate rhetorical heritage. Moreover, there might have been a connection between Heideggerian and Nazi thought if the Nazis meant what they said, i.e. about building a society in an age of nihilism.[46]

This, of course, is speculative, but, if what I have said is true, it seems that Heidegger's late work contains nothing less than a critique of Nazism. To be sure, that critique is not as explicit as his support for the Hitler regime from 1933 to 1935 but it is nonetheless there. I am suggesting that like his friend Ernst Jünger, Heidegger came to see that, despite all the propaganda about a new and regenerated society, the Nazis stood for just the opposite of their rhetoric.[47] If Heidegger attacked this movement, he did so in a manner appropriate to a metaphysician, not openly through political action in the conventinal sense, but in the classroom in his lectures on Nietzsche between 1936 and 1946.[48] This thesis is supported by the fact that there is hardly a 'philosophical' element in Nazi rhetoric which is not explicitly rejected in these lectures. Their subject is the figure whom the Nazis counted as their philosophical precursor par excellence, Friedrich Nietzsche. Yet Heidegger makes great efforts to prove that Nietzsche is first and foremost the prophet of the end of Platonism rather than any sort of political thinker.[49] In this respect he puts himself at odds with Alfred Bäumler, one of the most important Nazi ideologues. Heidegger insists that Nietzsche was neither fatalistic[50] nor biologistic[51] in his philosophizing but a metaphysician throughout. As much as Hitler and his coterie could adulate Richard Wagner, Heidegger explicitly asserts that Wagner and Schopenhauer, Wagner's philosophical mentor, were responsible for the dissemination of disastrous confusions such as a reified notion of will and an emotive theory of aesthetic value which ultimately lead to serious misunderstandings of the notions of action and evaluation not to mention the texts of Nietzsche.[52] Indeed, Heidegger explicitly defends and attributes to Nietzsche an intellectualistic and quietistic position under the rubric of "will to power".[53] He construes the latter as nothing more than self-overcoming and not at all a matter of force. As much as the Nazis emphasized the importance of *Weltanschauung*, Heidegger insists, as we have seen, that world views are a symptom of the failure of modern men to respond courageously to their paradigmless situation, instead forcing reality into a mould of their own making.[54] But what does all this have to do with the language of edification?

It seems that the state of paradigmlessness is a state in which we have lost all rational parameters for knowing and acting. In discussing the rôle of paradigms in science Kuhn writes, "To reject a paradigm without simultaneously substituting another is to reject science itself".[55] Now Heidegger's scenario is rather different from Kuhn's in that the paradigm which Platonism has supplied has dissolved rather than having been rejected but the net effect is the same; we are no longer in a position to ask for answers to the sorts of ultimate questions which have been the concern of traditional metaphysics the most important of which for social life being the legitimation of moral rules. We have seen that there are two basic responses to this condition: (1) arbitrarily imposing meaningful totality upon a fragmanted reality where there is none, i.e. forming a 'world view', or (2) accepting the way things are by following the path of thinking, which is essentially re-examination of the ultimate prerequisites for intelligibility and rationality purely and simply. The mystical turn in Heidegger's thought is connected with his efforts to lead us to rethink the elements upon which all thought rests. This obviously cannot be expressed in any but the most curious language because his aim is nothing less than inducing us to reflect upon what is pre-reflective, to articulate the unarticulated basis of discourse and to grasp what is pre-conceptual. In the later Heidegger this is called *das Ereignis* (the event of appropriation) the awareness that 'it gives' Being, the prerequisite of all possible horizons: language itself.[56] If this awareness is at all possible it must be on the basis of our capacity to get a glimpse of ourselves thinking about thinking rather than about things. But we have seen that Heidegger does not consider that this is possible in any literal sense; rather, the simple truth, *das Ereignis ereignet*, must be *meditated* upon. In short, we must *move* from the literal triviality of the expression to exploring its profundity. The language he employs to induce this movement will show what it says to those who have the capacity to follow its *Satz*. I have purposely used the German expression here, for Heidegger is well known to pun upon all of the meanings of the term *Satz*, sentence or proposition, leap and, last but by no means least, movement of a musical composition e.g. Schubert's c minor Quartetsatz. My claim, then, is that Heidegger's later thought aims at being musical by putting all of these senses together in his writing thereby utilizing musically structured sentences to induce a conceptual leap, a change in key, in his hearer. To understand what I mean by that, it will be necessary to look at the tradition of spiritual exercises from Ignatius Loyola to Kierkegaard and at the kind of language they incorporate.

The function of spiritual exercises is to create a spiritual disposition which will make it possible to choose to commit oneself to the Christian life, i.e. to move the person doing the exercises from a state where he is not only in doubt but lacking in procedures for confronting the resulting *anomie* to one where doubt is overcome and commitment is realized. It is a technique for inner conversion through affecting a change of heart, which in its turn makes possible a change of mind which is not merely an abstract change of opinion but a transformation of both one's way of living and one's outlook on life. By meditating upon the incompatibility of doubt and action, I come to realize that I cannot accomplish anything without putting my trust in something. This then becomes the stepping stone for making an act of religious faith.

The spiritual exercise proceeds from the assumption that true conversion must be a matter of the whole person not merely an affair of the intellect. The philosophical significance of such a step seems implicit in Plato's *Gorgias* in which it is argued that one must be converted from mysologistic egoism to philosophy, i.e. that there is a limit to self assertion, namely logic.[57] It is impossible for me to find satisfaction by doing *anything* I want to do because, if the only rule I accept is the rule that I ought to be arbitrary, then I never have any reason for wanting to do anything. I cannot formulate any plans according to this rule; so, if I do formulate courses of action as we all must, this will have to be on the basis of some other rule. I must recognize something other than my own self willing, something objective. But that means that my original rule has to be called into question. The only reason I would want to be arbitrary would be that it would be seen as good for me to do so. If it is not, I clearly would have no interest in so doing. Plato's point is a conceptual one based upon the notion that self-interest and a-fortiori the notion of the self does not make any sense unless that self recognizes that it is limited. Plato takes the recognition of this limit to be a kind of conversion to rationality or the point at which self-interest becomes enlightened. In Plato enlightenment is basically a commitment to seek reality objectively, which is to say, with the assistance of logic. The counterpart to the Sophists' egocentric ethics in epistemology, Plato recognized, is scepticism concerning the legitimacy of any rules whatsoever. Thus, the Sophist maintains that there are no moral rules. The snag is that this is supposed to be at least a hint, if not an actual guide, to right living — a moral rule. Just as pure selfishness logically pursued leads to a situation in which action and therefore pursuit of self interest is impossible, so the effort to deny that there are rules ends in establishing one. Plato's point is that there is a very important connection between selfhood and consistency such that, if you really think about what it is to be a self, you realize that it is impossible to be concerned with your self and totally arbitrary in you actions. Thus, concentrating upon what it is to be entirely arbitrary only show both that this is impossible and thus lead him who would consider the matter seriously to a conversion to philosophy. Without such a conversion to rationality *theoria* is impossible according to Plato. Once it has occurred *theoria* becomes a mode of self-construction. Philosophical discussion is fruitful only as long as everyone who is a party to that discussion is committed to rationality, i.e. examining the reasons people give for their views.

Spiritual exercises have a long history within and without Christianity before Ignatius Loyola produced his famous *Spiritual Exercises*, which made such practices preliminary to both Christian living and the study of philosophy and theology.[58] Loyola's exercises unite the Platonic insight with earlier traditions of pious exercises into a powerful program for transforming an individual's life by means of a four week training program. The aim of this program is to create a self-induced openness to choosing the Christian life through a series of dialogues which aim at creating a relationship between the individual and God. In the first week the literal meaning of the text is explored by the director who is essentially in dialogue with Loyola, the author of the text. In the second week the director and the individual making the exercise explore the deeper meaning, the formal significance, of the letter of the text. At the end of that period the individual is asked to make a choice for Christ. From that point forward the dialogue is between the individual and God. In the third week the individual explores the allegorical sense of the questions and answers that constitute Loyola's text. In the final week he learns God's will for him in an anagogical reading of the text. Thus a relationship is forged between God and the individual on the basis of the new language, which is continually being created in the course of the exercises. It is difficult to overstress the role of the spoken word in this process of reformation. The practice is less mystical than it is rhetorical. Unlike the Devotio Moderna and John of the Cross, Ignatius does not base his 'method' on images. The self, for example, is never described. What is produced is a scenario, rather than a vision. The individual learns to create a metalanguage, which is without an object language, through repetition. His own words conjure up cyphers which in their turn become constitutive of his character as the one doing the exercises 'introjects' the scenario he has created. It is important to point out that this is almost entirely a matter of auto-suggestion, in which the director simply guides the postulant without teaching him anything. The effectiveness of such an exercise is almost entirely a matter of the language in which it is cast.

The most important direct link between Heidegger and this tradition is statement that the most philosophically significant of Kierkegaard's writings, apart from the *Concept of Dread*, were the religious discourses.[59] It is not altogether clear what Heidegger was getting at in making this assertion; however, we do know that he unqualifiedly rejects Kierkegaard's Christianity with its theology of sin and its ontology of entities. What, then, is left to admire in Kierkegaard's discourses? The answer must lie in the style. And, indeed, the discourses are masterpeices of Danish prose. Their place in Kierkegaard's oeuvre is especially noteworthy.[60] They are ethical writings which aim not at eliciting the religious potential, by which Kierkegaard, well read in his Plato, understood coming to an awareness that to have real self is only possible to the extent that a person ceases to live *selfishly*. The goal of ethical discourse, then, is to affect that concern with the self that allows Plato's point about the construction of the self to show itself concretely within the life of the individual to whom they are addressed. In short, they aim at eliciting a sense of responsibility in that individual. As strange as this may seem to, say, a Rawlsian, it is altogether consonant with Kierkegaard's conception of the ethical as in some sense propaedeutic to the religious. Further, they are written entirely within secular categories, i.e. there are no appeals to revealed truths. Moreover, properly religious discourse, sermons, required the authority of the church, which SK did not possess in 1843 when he wrote the

Discourses. So he describes them as religious, rather than Christian, discourses. Just as the aesthetic writings provide brilliant 'phenomenological' descriptions of the empty inner life of the sensualist, so the religious discourses are no less brilliant efforts at edification, i.e. constituting a human being as a choosing subject.[61] On Loyola's scheme they endeavor to take the subject up to the point of where he must make a choice.[62] The distinguishing characteristic of Kierkegaard's spiritual exercises is the absolutely extraordinary way in which the language in which they are written accomplishes this.

We have seen that the aim of spiritual exercises is to induce movement or to persuade as well as to induce an experience of our temporally bound finitude as it contrasts with God's infinity. For Loyola this involved utilizing all the resources of classical rhetoric; for Kierkegaard, ever the master of dialectics, it meant nothing less than creating a new language, one which moved and, in the tempo of its movement, swayed our imagination. To accomplish this Kierkegaard insisted that his discourses be read aloud for their sound is as important as their substance. It is above all a musical sound. Kierkegaard's supreme effort in these discourses was to open his hearers heart and mind to the Living Word. In order to do this he had to make his words live i.e. do what they proclaim. He accomplished this by building tension into his discourses by creating fugal structures within classically balanced sentences. Repetition, alliteration and assonance, especially between cognate words, is the basis of his musical language. Thus in a single paragraph[63] SK introduces a subject with five motifs, produces variations on the motifs, inverts them and returns to the initial negative motif with which he began. In another place one can almost hear the crescendo as a section of a discourse moves to its conclusion:

> Herren gav; det er et kort Ord, men for Job betegnede det saa saare Meget; thi Jobs Hukommelse var ikke saaledes kort, og hans Taknemlighed ikke glemsom. Da hvlide Taknemligheden i hans Sjel med sit stille Veemod, han tog en mild og venlig Afskeed fra det Alt sammen, og i denne og i denne Afskeed forsvandt det Alt som en skjon Erindring, ja det syntes, som var det ikke Herren, der tog det, men Job, der gav ham det tilbage. Idet derfor Job havde sagt, Herren gav, da var hans Sind vel forberedt til at taekkes Gud ogsaa ved det naeste Ord: Herren tog.[64]

Unlike the treatise form in which one proceeds from a hypothesis to the evidence and principles of reasoning which establish that hypothesis, the structure of the discourses is not linear but circular. The discourses are "concentric circles", which are essentially musical in structure. There is typically an introduction, a meditation and two or three dialectical 'arguments'. The latter always lead back to the text which is the point of departure as well as a conclusion, which is a kind of coda. Further, to accommodate the various dramatis personae of the discourses — God, the scriptural writer, Kierkegaard and the hearer himself — he created a polyphonic structure similar to that of a motet. Kierkegaard modulates the flow of his discourse by means of an elaborate system of punctuation thereby creating a flow to time which urges and sways the hearer forward towards a decision.

While it is scarcely possible to catch the full richness of all of this in translation the German text which Heidegger would have read retains most of the crucial punctuation and therefore the solemnity of the original:

> Der Herr hat's gegeben, ist ein kurzes Wort, doch für Hiob bezeichnete es so viel; denn Hiobs gedächtnis war nicht so kurz, und seine Dankbarkeit nicht vergesslich. Da ruhte die Dankbarkeit in seiner Seele mit ihrer stillen Wehmut; er nahm einen milden und freundlichen Abschied von allem zusammen, und in diesem Abschied verschwand das alles wie eine schöne Erinnerung, ja es schien, als wäre es nicht der Herr, der es nahm, sondern Hiob, der es Ihm zurückgab. Indem deshalb Hiob gesagt hatte, der Herr hat's gegeben, war sein Sinn wohl vorbereitet, Gott auch mit dem nächsten Wort zu gefallen: "Der Herr hat's genommen".[65]

The antitheses and symmetry which produce the movement are barely affected by the translation.

The aim of all of this is nothing less than an act of self-creation. Herein lies the importance of such notions for Heidegger. For Loyola the spiritual exercises take place in an institutional context. Kierkegaard's genius radically transformed them into something entirely personal. Heidegger secularized meditation into dialogues between himself, an author such as Hebel or Trakl, language itself and 'that being which is always me' in much the same way as Kierkegaard creates dialogue between God, the apostle, himself and his hearer. This is appropriate in our state of paradigmlessness, for it is such that we have nothing to fall back on. It is not merely that we do not have the answers to our questions; we have no frameworks for posing the questions. We have to create both in a Münchausenesque act of extricating ourselves from the mire of paradigmlessness by our own bootstraps. All questioning depends upon substantive assumptions as R. G. Collingwood was wont to remind us. In the absence of 'eternal verities' we have only language, the "house of Being," through which we can create ourselves. In a sense language is the ultimate presupposition, for it is only through language that beings have their reality. However, language is not a substantive presupposition at all but that which makes creation possible, in Heidegger's terms the ultimate horizon.

His interest in Hebel and also several other masters of language, arises from the way in which they *show* the capacity of language to create from nothing as it were. Certain poets and philosophers (*Dichter und Denker*) have the capacity to lay bare the way in which language is the ultimate presupposition, the font of creativity. In short, they can exhibit the way in which language "gives" Being. To see his point we need simply ask ourselves the question. What is there that you cannot name? Thus, Heidegger harkens back to the origins of philosophy in Anaximander, Heraclitus and Parmenides to understand the intimate link between language and reality which Platonism obscured. From them he derives his fascination with the vocabulary of *Sein* (*einai*), *Wesen* (*ousia*), *Wahrheit* (*aleitheia*) and the like. He is also deeply concerned with the thought and the language of Meister Eckhart, the creator of the German philosophical vocabulary, who plumbed the depths of the presuppositions of Being itself. Eckhart is the source of several crucial terms in the later Heidegger's vocabulary; these include *Abgeschiedenheit*, the state of separation or detachment of the thinker who follows the path of thinking, *Gelassenheit*, the suspension of willing (and a fortiori rational thinking for Heidegger), the *Ort*, the place in Eckhart where the soul meets God and the origin of Being in Heidgger, as well as the notion of *Sammeln*, the collecting of things such that they become available.[66] It is not that Heidegger wants to return to an idealized past in turning to these thinkers in the era of paradigmlessness but that our problem is the same as theirs. We too have to begin anew with only the help of language itself. Abraham a Santa Clara, Hebel, Trakl and other poets show us in their works how language operates. Unless I am mistaken, what they show us is that language is what the gifted poet makes it to be. He is bound neither by conventional signification nor syntax. Thus, Abraham makes a moral point out of pure sound. Hebel creates a world in the *Alemanisch* dialect paying scant attention to the 'rules' of *Hochdeutsch* grammar. Trakl almost wholly disregards grammar to the point where it is virtually impossible to discern which adjective goes with which nouns or which nouns go with which verbs. Trakl creates a world out of ambiguity. In the words of one commentator:

> careful analysis of . . . any Trakl poem reveals a strong inner movement and counter-movement, evident in the rhythm and syntactic structure, the constellation of colors and vocal sounds and the change of tenses

and,

> the paucity of causal links, the interruptions of images and the metonomy have become exteme. Within the psychic space-time of the poem, all boundaries have been effaced in a radical manner.[67]

The point of the whole exercise in Heidegger's treatment of these figures is to make us aware that language is poetic usage. Meaning, rules of syntax and the like are all conventions. Language is poetic practice by means of which the world is created. With this in mind let us take a second look at the Hebel lecture.

It is closer to the sort of things we have seen in spiritual exercises than meets the eye. For example there is a pretty clear cut movement from simple fact to exhortation to meditation to revelation, which roughly recapitulates the four stages of the Ignatian exercises. Heidegger begins with a seemingly simple question: Who is J. P. Hebel? What follows clearly shows us that its simplicity and familiarity conceals nearly everything important about the subject. Heidegger answers this first question at a factual level.[68] Realizing that Hebel is not what he seems, i.e. a provincial or a writer of bad German (dialect), leads us to the question of just what Hebel actually is. Heidegger's response to this second question is startling, for he responds that Hebel is the *Hausfreund* (the early almanac-type calendar). Immediately the discourse takes on a new level as Heidegger unpacks the meaning of his metaphor.[69] His language becomes more extravagant as he begins to play: "So schränkte er den Schatz auf das Kostbare ein, baute ihm ein Schränklein und schenkte es im Jahre 1811 der ganzen Deutschen Sprachwelt als 'Schatzkästlein'".[70] The section concludes that Hebel's prose in the *Schatzkästlein* transmogrified the dialect poems of Hebel's earlier *Alemanische Gedichte*: "hinaufheben, verklären, veredeln und dadurch: verwandeln".[71] For a third time Heidegger asks who this person is, again answering that he is the *Hausfreund*, but now asking what we mean by *Haus* and *Freund*. Again the level of the discourse becomes more remote from everyday factual language as he explores the depths of yet more metaphors. The language of Eckhart enters into the picture when Heidegger speaks of Hebel as a house builder, i.e. a world builder, who constructs individual houses, "die in sich und um sich jenes vielfältige Zwischen *versammeln*".[72] Hebel, we now learn, is a preacher, not in the ecclesiastical sense but in the sense of one who speaks before or announces something. The subject of this announcing turns out to be nothing more than the nature stripped of the characteristics that calculating reason has concealed it with, nature in all its naturalness as *zuhanden* rather than *vorhanden*. We are now ready to understand who Hebel really is. In the fourth section of the lecture Heidegger finally reveals who Hebel is by explaing what he stands for metaphysically. The Hebel who is *Freund* of the *Haus* does nothing less than, "das Haus der Welt fuer das Wohnen der Menschen zur Sprache bringt."[73] He reveals the nature of language to us by leading us back to the unarticulated and unarticulatible pressupposition which underlies language. He can do this because he is close to the concrete spoken word, to his dialect. He can speak because he knows how to listen to language. Thus Heidegger comes to the dénoument of the lecture: "Eigentlich spricht die Sprache, nicht der Mensch. Der Mensch spricht erst, insofern er jeweils der Sprache ent-spricht".[74] Understood in the context of a spiritual exercise we thus find Heidegger striving after nothing other than a unity of word and deed, which will lead us indirectly via meditation to the very essence of language, linguistic action. On Heidegger's view this is an utterly simple notion and, as Aristotle pointed out long ago, one of the most difficult things for humans, with their complicated knowing apparatus to comprehend.

Heidegger stresses two closely linked aspects of Hebel's writing; Hebel's relationship to *Alemanisch* dialect and the other is his simplicity. The emphasis from the very beginning of Heidegger's talk is on the fact that it is very easy to mistake Hebel's language for bad German when it is actually superbly composed dialect. Heidegger is making two points here one specifically about Hebel and another more general one about the outlook most of us have on our environment. Concerning Hebel, he want to remind us that dialect is living language, which is not only the equal of polite *Hochdeutsch* but its superior to the extent that it is the language of simple ordinary people. The second point is a main element in all of Heidegger's philosophizing and a point that was equally dear to Ludwig Wittgenstein. It is that we have a tendency to take our environment too much for granted, to fail to see the splendor and mystery, the wonder of the everyday. We have a tendency to think of what is really important and profound as somehow hidden or remote when everything that we might hope to discover is in plain view all the time. It takes a poet to show us this:

Was wir gewönlich sehen von der Welt, von den menschlichen und göttlichen Dingen wird wir durch das dichterische Sagen in das Kostbare und in den überfluss des Geheimnisvolle umgeprägt. Das umgeprägende Veredeln geschiet durch eine gesteigerte Sprache. Aber die Steigern geht ins Einfache.[75]

Both in his poetry and his prose Hebel is always close to the way people speak. His 'erudition' is entirely ordinary and accessible to everyone who has given up the notion that wisdom is something esoteric. He has no doctrine but "er lässt den Leser gewärhen, damit dieser von sich aus in jener Zuneigung zum Wesenhaften gelange, zu dem sich der Hausfreund vorneigt, um uns zu sprechen".[76] It is both his and Heidegger's technique to lead us to insight rather than to articulate insights; they are both preachers of a sort — we must take our cues for understanding Heidegger from what he says about Hebel.

Hebel's mastery of ordinary speech becomes an avenue through which we gain access to nature in all its naturalness. Heidegger praises Hebel, citing Goethe, for giving us access to nature as *landscape*:

Der Verfasser verwandelt die Naturgegenstände zu Landleuten und verbauert, auf das naiveste, anmutigste Weise, durchaus das Universum; so dass die Landschaft in der man denn doch den Landmann immer erblickt mit ihm in unserer erhöhten und erheiterten Phantasie nur eines auszumachen scheint.[77]

The experience of landscape thus becomes the perfect antidote to the otherwise all pervasive effort to control nature. In short, Hebel is one of a small group of thinkers who have the ability to help us to confront our paradigmless situation not by trying to oppose it — this, Heidegger insists, is foolish, for its growth in our time is yet another aspect of the end of Platonism and in the long run just another way of becoming a servant to technology[78] — but by making us aware of the intimate link between language and reality. That awareness will be the prerequisite to transcending paradigmlessness to the extent that it is possible to do so. It is difficult to overstress the role that the cultivation of simplicity plays in attaining this self-awareness. It is in this context that what seem like *völkish* notions enter into Heidegger's thinking, for he identifies the splendor of the simple with landscape.[79] However, for one who has followed Heidegger's philosophical development the kinds of things he wants to say here cannot be taken in their literal sense and, what is more, there are good reasons why this must be so. This in itself would help to distinguish him from the tediously unimaginative Nazi and proto-Fascist writers. The later Heidegger may be replete with references to *Boden* and the family of conceptually realted terms but the *Blut* part of that infamous pair is totally lacking. The *Ort*, Messkirch, is the thinker's *Landschaft* but its values are simplicity and releasement (*Gelassenheit*). These are hardly the values of the Nazi or Fascist. In the final section of this paper I want briefly to indicate another current in German intellectual life to which we may look to find a similar set of values, one with which we can be sure Heidegger was familiar. I am referring to the Innsbruck periodical *Der Brenner* and the group around its editor Ludwig von Ficker.

Heidegger's lifelong relationship to *Der Brenner* and its editor is prima facie evidence for the kind of reading of his texts which I have suggested above.[80] Heidegger was a reader of the Brenner from 1911, the year after its inception and a point in the middle of what Heidegger himself has termed the most crucial phase in his intellectual development,[81] till it ceased publication in 1954. In 1952, on the occasion of his Trakl lecture at Bühlerhöhe, he seems to have invited Ludwig von Ficker to be in attendence. In any case, having met at that time, they became close friends and remained so till Ficker's death in 1967. There are unsubstantiated rumors that Heidegger considered Ficker something like a father confessor. Why should a world famous philosopher be so concerned with a provincial journal? The most obvious answer is that Ficker was the first to publish the magnificent lyric poetry of Heidegger's beloved Georg Trakl. Further, Ficker was one of the few friends that the tormented poet had. However, Heidegger's concern with the *Brenner* antedates Trakl's first contribution, which was submitted in 1912. So Trakl could not have been the initial attraction. What that attraction was remains to be discovered. However, we do know exactly what the early *Brenner* stood for. Even if we do not know just what drew him to the publication, a brief summary of its mission will help to illuminate the kind of issues which impressed the young Heidegger.

Originally Ficker, an aspiring actor and writer, was associated with a group of German nationalist writers collectively known as *Jung Tirol*.[82]

As *Jung Tirol* became ever more embroiled in the Tirolese *Kulturkampf* and, consequently, increasingly anti-Catholic it struck Ficker among others that the German nationalists in Tirol had exchanged on set of obnoxious ideological blinkers, establishment Roman Catholicism, for an equally obnoxious set of Pan-German ideological blinkers when what was required was a movement to transcend ideological blinkers altogether. In this group he encountered a like-minded individual, Carl Dallago, who had cast off Catholicism but who found the alternatives to it no less legalisitc, moralistic, conformist and confining. The *Brenner* was founded principally to give voice to Dallago's assault upon dogmatism.[83]

Having observed the clash of ideologies around him, Dallago inferred that the fundamental problem in his society was the intellectualization of values. People spent so much time developing moral, social and aesthetic theories that they failed to reflect upon just what it was to live as a decent human being. Like the Viennese writer, Karl Kraus, whom he admired (and who returned his admiration), Dallago was convinced that the arguments of those who spoke in the name of morality were nothing more than instruments designed to force dissenters into conformity. The counter-arguments of the liberal nationalist establishment were no less manipulative for being advanced under the aegis of natural science and progress. On Dallago's view of the matter both parties went wrong when they tried to equate human decency with holding to a particular theory of right and wrong. The parallel ways in which both groups erred convinced him that reason was essentially calculating and manipulative. If one were really to be a decent human being, one would have to reject thought for life. He took this to be the authentic teaching of philosophy (Socrates), of religion (Jesus) and even of science (Giordano Bruno). The quest for certainty in intellectual matters (the counterpart of the quest for material security) has made philosophers, theologians and scientists conceive what there is as sharply defined, abstract and static, i.e. as something wholly unlike the 'booming buzzing confusion' the ordinary person encounters. Above all, the really real of the philosophers, scientists and theologians is coldly asexual. For Dallago this was enough to indicate that the cultivation of intellectual values was eo ipso corrupting. That this entailed that he reject everything that urban civilization stood for did not bother Dallago in the least; for he was firmly convinced that only someone who had experienced life close to the landscape could make this, essentially religious, discovery. His rejection of urban values, however, must be distinguished from that of *völkisch* thinkers in that Dallago's love for nature has nothing whatsoever to do with any sort of peasant ideology. He in no way glorifies the peasantry; in fact he never even discusses such things. Rather, his writings — essays, poems, aphorisms — celebrate and exhort his reader to experience the mysteries of the ever changing landscape. Unlike the proto-Fascists, this is what he found in Nietzsche, who, along with Jean-Francois Millet, Giovanni Segantini and Walt Whitman, was his great hero. In Whitman's case this took not a little courage on Dallago's part, for, after having been all the rage in the 90s and the first years of our century in the German speaking world, Whitman had become scorned after the discovery of his homosexuality. If we can speak of an ideal life according to Dallago, it would have to be a solitary life close to the landscape, abjuring security, and celebrating the wonder of nature's mysterious rhythms poetically. These were the qualities which intellectuals like Max Brod (who was sufficiently impressed with Dallago to introduce him to Franz Kafka) admired in Dallago and the values which gave the *Brenner* its orientation. The parallels to the later Heidegger hardly need drawing at this stage.

In the course of its forty four year development the *Brenner* had basically two foci: both were rooted in Dallago's program, one which stressed the importance of the simple life and, conversely, excoriated everything decadently conformist and another which emphasized poetic creativity and the worth of the individual personality. The journal, then, was basically dedicated to literature and a kind of social criticism which emphasized the regeneration of a corrupt society through a proper, anti-ideological, view of the world. To the extent that he was able to its editor, Ludwig von Ficker, strove to make the *Brenner* as open as possible to contemporary literary and ethico-religious currents, provided that the latter were committed to tolerance. His deep concern for linguistic innovation led Ficker to publish the Expressionist poet Theodor Däubler in an early number of the *Brenner*. However, Ficker's discovery of Georg Trakl must count as his greatest contribution to German literature. Moreover, in it effort to emulate and to champion that morally based satire of Karl Kraus, the *Brenner* was closely associated with the great Viennese critique of language, which Kraus carried on in the pages of his periodical *Die Fackel*.

In his efforts to explore new ways of thinking about the individual and society Ficker came to publish the work of two philosophers, Theodor Haecker and Ferdinand Ebner, as well as Haecker's translations of and commentaries on the works of Søren Kierkegaard. To be sure Haecker and Ebner were thinkers who had much less in common with Heidegger than Dallago; but Heidegger's continued reading of the publication in which they published must be indicative of some common concerns.[84] Moreover, since all of the *Brenner* philosophers were opposed to everything the Nazis stood for — Haecker was among the first Munich intellectuals arrested by the Nazis and Ebner was a model for the Catholic socialist among the resistance — the very fact that Heidegger was a lifelong subscriber is a significant datum for assessing a thesis like Minder's.

The central notion in Ebner's thought is that language and human personality are intimately bound together to the extent that and I and a Thou only encounter each other in speech (and conversely speech, as opposed to written language is the key to understanding language). Ebner was every bit as much committed to the primacy of the practical over the theoretical as Heidegger. Haecker's odyssey from Protestant champion of Kierkegaard to heterodox Catholic theologian (whose concept of Christian culture strongly influenced *inter alia* T. S. Eliot) did not prevent him from attacking the Nazi's at the time of their seizure of power in Germany in the pages of the *Brenner*. Indeed, the virulence of Haecker's anti-Nazi polemic, which condemns secular society as "murderous, plebian and Protestant"[85] seems to be only slightly less fanatical than that of the Nazis he opposed. Whatever Heidegger's opinions of these and other notions of Brenner authors, it is surely significant that it was in the pages of the *Brenner* that the first German editions of Kierkegaard's religious discourses appeared in Haecker's translations. There was, then, not a little to interest Heidegger in the pages of the *Brenner*. Indeed, in one form or other most of his fundamental concerns, especially in his later writings, are present in these anti-völkisch thinkers. So his association with the *Brenner* should not pass without comment in any discussion of his intellectual development, for it clearly provides us with examples of thinkers who share many of Heidegger's allegedly *völkisch* beliefs but who were clearly anti-*völkisch*.

It is time to draw all of this together. What kind of relationship exists between Heidegger's late style and Nazi rhetoric? The best answer seems to be that they share certain family resemblances. However, they should not be equated on that account, for there are other instances of such a fallacy in the intellectual history of Europe. For example, John Boyer has argued persuasively that there has been a tendency to run together Baroque paternalisitc authoritarian rhetoric and Fascist totalitarian rhetoric without justification.[86] Boyer maintains that we cannot understand the political vocabulary of a Karl Lueger if we do so. These rhetorical modes may be similar and may even converge but they are not to be identified under pain of anachronism. I believe that there is a complex process of descent from the Baroque sermon to Nazi rhetoric which is a genuine evolution in that it is branching rather than a continual line of uninterrupted development. This accounts for the similarities but by no means allows us to make the kind of identity claims that Minder presses. Moreover, in the most brilliant book yet to be written on Hitler Peter Stern has established that Hitler's successes in convincing people of the rightness of his cause where due to the fact that he was able to speak to his age in the language it understood best.[87] He makes a strong case for the claim that the essence of Hitler's power to convince lay in his ability to create what he promised, a sense of a united *Volk*, in those who listened to his interminable speeches. In short, his speeches were what John Austin termed perlocutionary acts. Now this is just what the language of spirituality,[88] and, if I am correct, what Heidegger aims at. But they are very different kinds of perlocutionary acts. Hitler and Co.'s acts, as Stern rightly maintains, were directed at precisely the kind of self-assertion which Plato decried as self-destructive. Hitler's end was conceptually linked to his one and only true belief, the faith in his own will, which turns out to be nothing other than arbitrariness. Heidegger, on the other hand, directs the perlocutionary force of his speech to a realization which rests upon analyses of the nature of action and communication, a view of the theory of language and an account of the development of western civilization, all of which are philosophically defensible when they are not part of the main stream of twentieth century analytical philosophy. My suggestion is that, after his withdrawal from politics,[89] Heidegger implicitly or explicitly understood how Hitler used the language of the age and began to counter that movement with his own version of it. He could understand and, for a time, identify with a movement that spoke his language for the same reasons that Ficker could be completely opposed to everything the Nazis stood for and, in his capacity as editor of the *Brenner*, have corresponded with the young Carl Schmitt and the young Alfred Baeumler both of whom became notorious under the Nazi regime.[90] What looked like shared concerns in both cases turned out to be very different enterprizes. If my view of Heidegger is correct, his response was too little, too late, politically and morally but that is to say nothing of its philosophical significance. The stylistic issue, then, is crucial for understanding and evaluating Heidegger; it is ignored at our peril but we must consider the full stylistic context, i.e. the literary elements must be taken within the context of the philosophical strategy they serve. Stylistic issues alone can never be a substitute for those philosophical considerations which alone are relevant to the appraisal of a philosophical position.

NOTES

* I am indebted to Ken Winkler and David Kolb for illuminating comments on this chapter.

[1] Robert Minder, "Heidegger und Hebel oder Die Sprache von Messkirch", *Dichter in der Gesellschaft* (Frankfurt/Main, 1966), pp. 211—64.

[2] Paul Hühnerfeld, *In Sachen Heidegger* (Hamburg, 1959), p. 85.

[3] Martin Heidegger, *Hebel der Hausfreund* (Pfullingen, 1957), p. 26.

[4] George Mosse, *The Crisis of German Ideology* (New York,1964), *passim.*

[5] For an historically accurate account of Hebel and his work see Uli Daester, *Johann Peter Hebel in Selbstzeugnissen und Bilddokumenten* (Reinbeck, 1973).

[6] Wilhelm Shäfer, cited by Minder, "Heidegger und Hebel", p. 215.

[7] Heidegger, "Der Feldweg", *Wort und Wahrheit* **5** (1950), p. 267.

[8] Heidegger, "Der Spruch des Anaximander", *Holzwege* (Frankfurt/Main, 1950), pp. 296—343; "Die Sprache im Gedicht", *Unterwegs zur Sprache* (Pfullingen, 1959), pp. 35—82.

[9] Personal communication from Norbert R. Wolff, professor of German, Würzburg University.

[10] Interview with Martin Heidegger, "Nur ein Gott Kann uns Retten", *Der Spiegel* **23** (1976).

[11] Heidegger, *Sein und Zeit* (Tübingen, 1977), pp. 153—60; cf. "Zeit und Sein", *Zur Sache des Denkens* (Tübingen, 1976), pp. 24—5.

[12] George Steiner, *Heidegger* (London, 1978), pp. 112—21.Steiner's views on this matter, i.e. his reluctance to pronounce upon Heidegger one way or another, deserve to be taken especially seriously on account of his vast knowledge of German lanugage and literature and his commitment to the notion that the Nazi experience utterly transformed German; see his *Lanugage and Silence* (New York, 1967).

[13] Steiner, *Heidegger*, p. 113.

[14] If I am not mistaken, this is the point of Heidegger, *Hebel*, pp. 24—7.

[15] Anthony Kenny, *Wittgenstein* (Harmondsworth, 1973), pp. 219—32.

[16] R. G. Collingwood, *An Essay on Metaphysics* (Oxford, 1940), p. 32.

[17] This is emphasized by Lucien Goldmann in his *Lukács et Heidegger* (Paris, 1973), pp. 91—105.

[18] Minder, "Heidegger und Hebel", p. 231.

[19] Robert A. Kann, *A Study in Austrian Intellectual History* (New York, 1960), pp. 92—3.

[20] Abraham a Santa Clara, "Der alte Hafen scheppert", *Auswahl aus Abraham a Santa Clara*, ed. Karl Bertsche (Bonn, 1911), pp. 3—4.

[21] Abraham a Santa Clara "Hui and Pfui der Welt", p. 479, cited in Kann, p. 84.

[22] It is difficult to overemphasize the manifold ways in which the Nazis exploited both popular and high culture. In the course of a seminar given jointly by Bernhardt Blumenthal and me at LaSalle College in 1972 we began to notice the derivative charater of many Nazi songs. The researches of Vernon Lidtke also bear out this point. Personal communication. Their exploitation of the philosophical tradition for propaganda purposes (following in the tradition of the jingoists during World War One) was no less thorough. The case of their exploitation of Nietzsche has been discussed at length by Walter Kaufmann and others; it is but the best known of several.

[23] Those who feel squeamish with Heidegger's metaphysical language may overcome their misgivings about the history of Being by means of an analogy drawn from contemporary philosophy of science. I am referring to Popper's notion of World 3. Like Popper, Heidegger is a quasi-idealist in that he wishes to single out a universe of discourse (a term he would probably not like) peculiar to the history of ideas (a project that he shared with the other Dilthey enthusiasts of his generation) but without commiting himself to an ontology of beings such as that of medieval Christianity in which there are things such as being or Being. However, unlike Popper's World 3 the history of Being is not a set of propositions. The most difficult problem that Heidegger's language presents stems from the fact that he uses the vocabulary of traditional metaphysics and Pietist religious language to express notions that are in no obvious traditional sense metaphysical or religious.

[24] Ludwig Wittgenstein, *Vermischte Bemerkungen* (Franfurt/Main, 1977), p. 53.

[25] Eg. Wilhelm von Humboldt's notion of an "innere Sprachform" in *Schriften zur Sprachphilosophie, Werke* (5 vols.; Stuttgart, 1963), III, p. 192.

[26] Heidegger, *Sein*, pp. 66—76 *et passim.*

[27] Heidegger, *Sein*, pp. 2—14.

[28] Rudolf Carnap, "The Elimination of Metaphysics Through Logical Analysis of Language", *Logical Positivism*, ed. A. J. Ayer (Glencoe, Ill., 1959), p. 69.

[29] Heidegger's move in saying things like *das Nichts nichtet* is best understood by comparison with G. E. Moore's anti-reductionist use of the term good in the sentence 'the good is good'. Moore insisted that this is all one can say about the good but that the sentence cannot be a proposition (i.e. true or false). G. E. Moore *Principia Ethica* (Cambridge, 1971), pp. 1—17.

[30] Hühnerfeld has emphasized this, *In Sachen*, p. 75ff.

[31] Walter Sokel, *The Writer in Extremis* (Stanford, 1959), p. 3.

[32] Walter Kaufman, Prologue, Martin Buber, *I and Thou* (New York, 1970), pp. 7—48, 110n. 6; cf. *Discovering the Mind* (3 vols.; New York, 1979—1981), II, p. 206 *et passim.*

[33] Hühnerfeld, *In Sachen*, p. 115.

[34] Heidegger, *Sein*, p. 38.

[35] On Husserl see Peter Koestenbaum's introductory essay to his edition of Husserl, *The Paris Lectures* (The Hague, 1970), pp. ix—lxxvii.
[36] Heidegger, *Sein*, p. 161.
[37] I have explored some of these similarities in a lecture on Wittgenstein and Heidegger at the University of Tel-Aviv in 1979.
[38] Heidegger, *Sein*, pp. 175—80.
[39] Heidegger, *Sein*, pp. 164—65.
[40] Heidegger, *Sein*, pp. 153—60.
[41] I have discussed this in "Wittgenstein: an Austrian Enigma", *Austrian Philosophy*, ed. J. C. Nyíri (Munich, 1981), pp. 75—89.
[42] Thomas Kuhn, *The Structure of Scientific Revolutions* (Chicago, 1970), p. 103 *et passim*.
[43] Heidegger, *Nietzsche* (2 vols.; Pfullingen, 1961), I, pp. 253—38.
[44] Kuhn, *Structure*, p. 13.
[45] Heidegger, *Nietzsche*, II, p. 21ff.; cf. "Die Zeit des Weltbildes", *Holzwege*, pp. 69—104. It is worth contrasting Heidegger's view of the matter with Hitler's: "Alle diese Verfallserscheinungen sind im letzten Grunde nur Folgen des Mangels einer bestimmten, gleichmässig annerkannten Weltanschauung sowie der daraus sich ergebenden allgemeinen Unsicherheit in der Beurteilung und der Stellungnahme zu den einzelnen grossen Fragen der Zeit", *Mein Kampf* (Munich, 1938), p. 292.
[46] Nearly all authorities on the Third Reich are inclined not to take the 'official' Nazi ideology as representing what the Nazis really stood for. Hilter once remarked that he found Rosenberg too tedious to wade through. Hitler's jettisoning of Feder's 'socialist' element in the party to woo the industrialists and his elimination of the *alt Kämpfer* in the SA are two examples of failing to take ideology seriously. J. P. Stern suggests that the only element in the Nazi ideology to which we can be certain that Hitler adhered was belief in his own will, J. P. Stern, *Hitler* (Berkeley and Los Angeles), p. 61.
[47] Stern, *Ernst Jünger* (New Haven, 1953), p. 13—7 *et passim*.
[48] Hühnerfeld writes that Heidegger's students reported Heidegger complaining about the Nazis in 1936, *In Sachen*, p. 93.
[49] Heidegger, *Nietzsche*, I, pp. 29—33, p. 650.
[50] Heidegger, *Nietzsche*, I, p. 399.
[51] Heidegger, *Nietzsche*, I, p. 367.
[52] Heidegger, *Nietzsche*, I, p. 101ff.
[53] Heidegger, *Nietzsche*, I, p. 185 *et passim*.
[54] Heidegger, *Nietzsche*, II, p. 255.
[55] Kuhn, *Structure*, p. 79.
[56] Heidegger, "Zeit und Sein", pp. 21—5; cf. *Nietzsche*, II, p. 489.
[57] Plato, *Gorgias*, pp. 481b—527c; cf. *Epist.* VII, p. 340c ff. on the need to be 'converted' to philsosphy. Spiritual practices (from which spiritual excercises in the strict sense evolve) enter philosophy with the Pythagorean cultivation of puns (the classic is soma-sema body-tomb) which is nothing other cultivating seeing the necessary in the accidental. In the ancient world this reaches its zenith in the Stoics who developed truth functional logic as part of a technique for seeing *everything* as occurring necessarily. To the extent that Kantianism is an offshoot of Stoicism and foundationalism is essentially Kantian (regardless of what its proponents allege) this form of spiritual practice has had an enormous influence on modern thought. What

Heidegger calls Platonism (carefully distinguishing it from the philosophy of Plato) should be historically identified with Stoicism. Eduard Zeller's account of Stoicism brings this out in a way that modern treatments tend to gloss over. See Eduard Zeller, *The Stoics, Epicureans and Sceptics,* trans, Oswald J. Reichel (London, 1870), pp. 36—402. I have benefited from discussion of foundationalism with Owen Flanagan.

[58] On Loyola see Roland Barthes, "L'arbre de la foi", preface to Ignatius Loyola, *Exercises spirituels* (Paris, 1972), pp. 5—53.

[59] Heidegger, *Sein*, p. 235.

[60] See the superb bilingual (Danish-French) edition with introduction and full commentary, *Søren Kierkegaard, Hâte-toi d'ecouter*, ed. Nelly Viallaneix (Paris, 1970). That these works are at the the center of SK's thought is attested to in his statement, "I held out Either/Or to the world in my left hand and in my right the *Two Edifying Discourses* but all, or good as all, grasped with their right what I held in my left", *The Point of View for My Work as an Author*, trans, Walter Lowrie (New York, 1962), p. 20. It is not difficult to see a complex allusion to Lessing's parable in which God offers the choice of all truth in His right hand or the desire of all truth even though we err constantly choice in his left (Lessing would choose the left) in this text.

[61] See Andre Thébaut, "Edification", *Dictionnaire de spitituralité*, (Paris, 1937—), **IV**, pp. 287—93.

[62] It is enticing to conceive the whole Kierkegaardian oeuvre as constructed around the idea of Loyola's spiritual exercises. If it were, the aesthetic would correspond to the first week, the literal exploration of the text which leads to repentance for sin, the ethical would correspond to the moral lessons on takes from reflecting upon the life of Christ up to Palm Sunday and the point where amendment and reformation of the exercitant's life must be chosen, religion A would correspond to the third week when the passion is contemplated and the newly converted exercitant prays for divine illumination, and religion B would correspond to the fourth week in which the end is unity with the devine. One of SK's favorite pseudonyms, Johannes Climachus, i.e. that of a famous author of spiritual exercizes, who influenced Ignatius Loyola, see André Rayez, "Interiroité" *Dictionnaire de spiritualité*, **7**, pp. 1890—1904. The exact significance of this is uncertain.

[63] Kierkegaard, *Hâte-toi*, discours 2, par. 15.

[64] Kierkegaard, *Hâte-toi*, **I**, p. 6.

[65] Kierkegaard, *Religiöse Reden*, trans. Theodor Haecker (Leipzig, 1936), pp. 109—10.

[66] On Eckhart and Heidegger see John Caputo, "Meister Eckhart and the Late Heidegger", *Journal of the History of Philosophy* **12**, (1974); **13** (1975) 61—80; cf. Caputo's *The Mystical Element in Heidegger's Thought* (Athens, Ohio, 1978). I have benefited greatly from numerous conversations and discussions of Heidegger with Professor Caputo. I am also grateful to Professor Caputo for allowing me to read the typescript of his book. For an appreciation of Meister Eckhart's contribution to German vernacular philosophical vocabulary from a very different perspective (but nonetheless enthusiastic for all that) see Fritz Mauthner, *Wörterbuch der Philosophie* (2nd. ed.; 3 vols.; Leipzig, 1923—1924), **III**, pp. 371—79.

[67] Marie Jaarus Kurrik, *Georg Trakl* (New York, 1974), p. 29.

[68] Heidegger, *Hebel*, pp. 5—8.

[69] Heidegger, *Hebel*, pp. 8—12.

[70] Heidegger, *Hebel*, p. 11.
[71] Heidegger, *Hebel*, pp. 11—2.
[72] Heidegger, *Hebel*, p. 13. My emphasis.
[73] Heidegger, *Hebel*, p. 25.
[74] Heidegger, *Hebel*, p. 26.
[75] Heidegger, *Hebel*, p. 12.
[76] Heidegger, *Hebel*, p. 19.
[77] Goethe, cited by Heidegger, *Hebel*, p. 22.
[78] Heidegger, *Gelassenheit*, p. 24.
[79] This is clearest in the philosophical fable, *Der Feldweg*; see n. 7.
[80] I have discussed certain aspects of the relationship of Heidegger to *Der Brenner* in "Carl Dallago und Martin Heidegger", *Untersuchungen zum Brenner*, eds. Methlagl, Sauermann and Scheichl (Salzburg, 1981). Walter Methlagl has discussed the relationship of the late *Brenner* and Ficker personally to Heidegger (and T. W. Adorno) in "Die Zeit und die Stunde der Zeit", *Studien zur Literatur des 19. und 20. Jahrhunderts in Oesterreich*, eds. Holzner, Klein and Wiesmueller (Innsbruck, 1981), pp. 153—78. Heidegger's high estimate of Ficker is reflected in the fact that he spoke extemporaneously in his honor in 1960 when the Free University of Berlin conferred an honorary degree on the *Brenner's* editor. Heidegger's remarks on that occasion are recorded in a privately published booklet commemorating the occasion, *Ludwig von Ficker zum Gedächthnis seines achtzigsten Geburtstags*.
[81] Hans Seigfried, "Martin Heidegger: A Recollection", *Man and World* **3** (1970), pp. 3—4.
[82] The edition of Ficker's correspondence in preparation at the Brenner Archive in Innsbruck under Walter Methlagl's direction contains an account of the origins of the periodical. See also, Gerald Stieg, *Der Brenner und Die Fackel* (Salzburg, 1976) and my, "Carl Dallago and the Early *Brenner*," *Modern Austrian Literature* **11** (1978), pp. 1—17. I am grateful to Drs. Methlagl and Stieg for many conversations and suggestions about the *Brenner* and its relation to twentieth century philosophy.
[83] See n. 80.
[84] On Haecker see Eugen Blessing, *Theodor Haecker* (Nürnberg, 1959); on Ebner see Theodor Steinbuchl, *Der Umbruch des Denkens* (Regensburg, 1936). I am indebted to Friedrich Lehne for information about Ebner as a model for Catholic socialists.
[85] Theodor Haecker, "Betrachtungen über Vergil, Vater des Abendlandes", *Der Brenner*, **17** (1932), p. 30.
[86] John Boyer, *Political Radicalism in Imperial Vienna* (Chicago and London, 1981), p. 464.
[87] Stern, *Hitler*, p. 20 *et passim*.
[88] See Jean Ladrière, "Langage des spirituels", *Dictionnaire de spiritualité*, p. 9, pp. 204—17.
[89] See n. 48.
[90] The letters are in the Brenner Archive; they shall be included in the forthcoming edition of Ficker's correspondence.

CHAPTER II

NYÍRI ON THE CONSERVATISM OF WITTGENSTEIN'S LATER PHILOSOPHY

In his keynote address to the 2nd International Wittgenstein Symposium at Kirchberg am Wechsel in Lower Austria in 1977 G. H. von Wright announced the publication of Wittgenstein's *pensées* on philosophy, culture and society from Wittgenstein's *Nachlass* and argued that the publication of the *Vermischte Bemerkungen* pressed the question of the relationship between Wittgenstein's personal beliefs and his philosophical positions upon us more poignantly than even before.[1] Wittgenstein's enthusiastic endorsement of Oswald Spengler's *Kulturpessimismus* more than anything else made it imperative in Professor Wright's view that we determine whether the connection between Wittgenstein's personal and philosophical beliefs is merely historical and psychological, or logically and conceptually linked. I have long believed that there is unity between Wittgenstein's life and his thought. Indeed, I have made this view the cornerstone of my researches into Wittgenstein's intellectual and moral heritage. However, I have refrained from articulating the precise nature of this link for two reasons: on account of the paucity of reliable information about his life, on the one hand, and the remarkably difficult problem of establishing such logico-conceptual links between personal beliefs and philosophical positions, on the other. I, therefore welcome the bold efforts of the distinguished Hungarian philosopher, cultural historian and Wittgenstein scholar J. C. Nyíri to confront this complex and vexing problem.

In a series of papers Nyíri has argued that the tone, substance and historical context in which Wittgenstein's later work developed all clearly mark the *Philosophical Investigations* as a representative of what is best in a peculiar brand of conservative thought, one which Nyíri also seems to endorse himself. Drawing heavily on *Vermischte Bemerkungen*, which he regards as the manifesto in which Wittgenstein expressed his world-view, and a wealth of historical and philosophical information, with full awareness of the complexities that surround the issue; he has ventured the hypothesis that there is a relation of implication between the conservatism embodied in Wittgenstein's personal values and, specifically, his notion of "forms of life". Indeed, on Nyíri's account Wittgenstein "saved . . . the Neo-Conservative position from a theoretical catastrophe at a time when, in Germany, it could no longer be saved from a political catastrophe".[2] In order to establish this thesis Nyíri musters both historical-biographical data as well as conceptual-philosophical argumentation to establish (1) Wittgenstein's personal conservatism, (2) just what the notion of conservatism designates, in particular, the nature of Neo-Conservatism and (3) a conservative interpretation of such central notions in the *Philosophical Investigations* as rule-following and forms of life. Because I take Nyíri's first two points to be questionable and the third simply incorrect, I want to criticize the manifold aspects of his arguments with a view to laying further ground work for investigating the relationship between Wittgenstein's life and his thought. I must make it clear from the outset that the problems and difficulties I find in Nyíri's argument are for the most part subtle rather than gross errors. They are largely points of interpretation to which there is an alternative. No single point, when refuted, would be sufficient to force Nyíri to modify his claim. Thus, I must take issue with him point by point to establish how the argument rests upon shaky foundations. I do not think that his interpretations are implausible but that he has presented many things as though there was no evidence to the contrary or no difficulties in maintaining his position or interpretation. Consequently, I have chosen to resist his thesis by challenging it at as many junctures as possible, not because I have any great faith in nit-picking or mere quantity, but to emphasize alternative readings and evaluations of Wittgenstein, his work and conservatism. My hope is that this dose of skepticism will help to refine our understanding of the pressing and perplexing issues he has raised. I shall begin by discussing the problems surrounding Nyíri's use of the term "conservatism".

Both the concept of conservatism that Nyíri employs and what might be termed his "conservative mode of analysing" Wittgenstein (which he mistakely identifies with a Wittgensteinian approach) relies on assumptions that are at the very least less certain and obvious than he takes them to be. Briefly, I want to argue that:

(1) Nyíri's exposition of conservatism fails to take into consideration the plethora of ambiguities which surround the concept.

and, consequently,

(2) are such that the claim that an account of the nature of, say, thinking or knowing which is not explicitly "conservative" becomes difficult to assess.

This is because

(3) Nyíri has not recognized that there exist difficulties surrounding appeals to the notion of "family resemblances" in historical and sociological explanations.

and, finally, because

(4) his approach to Wittgenstein as a conservative involves appeal to a dubious (and non-Wittgensteinian) idealistic mode of explanation.

So, I want to challenge his account of conservatism, his appeal to the idea of the "family resemblance" character of the notion, its appropriateness for categorizing Wittgenstein's epistemology and the underlying assumption Nyíri makes about the relationship between belief, theoretical positions underlying critical practices and political activities. To do this I must first look at what Nyíri actually says about Wittgenstein and conservatism.

To begin with, he asserts, "the specific tone of Wittgenstein's analyses, the content of many of his remarks and reflections and the historical circumstances in which this philosophy came into being definitely invite an interpretation in the light of which there indeed emerges family resemblances between Wittgenstein on the one hand and some important representatives of conservatism on the other".[3] Nyíri is well aware that the present state of Wittgenstein's *Nachlass*, as well as our imperfect knowledge of Wittgenstein's biography makes it difficult to do more than *suggest* points of contact with conservatism. He is no less aware that conservatism is basically ambiguous but, I think that he is unaware of the extent of that ambiguity.

Further, his covert idealism, among other things, prevents him from carrying out an accurate assessment of the ambiguities surrounding the term. The idealism I allege to be characteristic of Nyíri's mode of explanation is apparent in the following lengthy passage:

> The term "conservative" is used in at least three types of context. People speak of a conservative attitude or mentality, of conservative theory and of conservative politics. Conservative attitude, theory and politics are of course by no means independent of each other. Conservative theory comes into being, in certain social and historical circumstances, as an abstract self-perception of conservative mentality, presenting the latter as the only acceptable or indeed normal one. Conservative theory can take the

form of anthropology, social theory or the theory of history: in its content it can embody very different tendencies, depending on what remains — if anything at all — that is regarded as worth conserving or even re-establishing. Conservative politics, finally, are upheld by conservative mentality, and directed by conservative theory — though this latter relationship is already a rather difficult one, since conservative mentality and thus also conservative politics have a distaste for any theory.[4]

I am not as certain as Nyíri is that such a relationship actually obtains between attitudes, theories and politics, and it is important that I mention that it is among other things Wittgenstein's *Philosophical Investigations* that has convinced me that this idealist approach to the history of ideas is dubious. This idealism I allege to be present in Nyíri's work consists precisely in assuming that ideas and theories (but also critical techniques) must have the significance that their discoverers and proponents claim for them. I am claiming that we are in no way bound to accept the conservative's claims to be the spokesman for tradition, that there is in fact a good deal of evidence to the contrary. Alasdair MacIntyre draws attention to this "false consciousness" in his *Short History of Ethics* where he writes, "A striking feature of moral and political argument in the modern world is the extent to which it is innovators, radicals and revolutionaries who revive old doctrines, while their conservative and reactionary opponents are the inventors of new ones".[5] Likewise, Edward Thompson has argued eloquently that the tradition of the Common Law has formed and substained English working class radicalism.[6] I am suggesting that the relationship Nyíri assumes to exist between conservative practice and conservative theory involves an appeal to the ethical and social counterpart to the literary scholar's "Intentional Fallacy". These are the bare bones of my argument, I must now put some flesh on them. My starting point will be the concept of conservatism.

Consider the following statement from the forward to one of the books Nyíri cites: "The liberal-minded among the conservative aristocracy were few".[7] The paradox it enunciates rests upon the fact that it seems to be a conceptual impossibility but, when said of the social attitudes of the 19th century *Junker's* attitudes to (paternalistic) social reform, it is an empirical fact that there were indeed liberal-minded conservative aristocrats. I suggest that Nyíri ought to perceive a problem here. Clinton Rossiter certainly does, in the face of such ambiguities, he opens an encyclopedia article on conservatism by asserting, "'Conservatism' is a word whose usefulness is matched only by its capacity to confuse, distort and irritate", it is, "a handy, if dangerous tool of social science".[8] Because this is so, Rossiter argues that social scientists who utilize the term have an obligation to establish criteria for their usage. Thus Rossiter devotes the remainder of his article to setting out guidelines for the use of the term conservative.

In fact, Nyíri's appeal to the Wittgensteinian notion of "family resemblance" to elucidate the concept of conservatism is question-begging. This is a serious charge, so I should clarify my claim. His appeal to the notion of family resemblance, far from absolving Nyíri from giving us an account of conservatism and neo-conservatism, requires him to justify his allegation explicitly. Further, to assert, as Nyíri rightly does, that the various phenomena commonly called "conservative" relate to each other as members of a family, who only share *some*, as opposed to all, characteristics in common, also implies *pace* Nyíri that "conservative" will *in some way* have to be defined by reference to the substance of the issue which is at stake. This is precisely because there is no absolute unity to the concept. To speak of general notions such as this as unifying clusters of characteristics, as opposed to the universality of, say, an Aristotelian definition, is to assert that such conceptions are open-textured, which, in turn, is to assert that it is a matter of debate just what falls under them — and that there is no guarantee that the debate can be resolved. When the scope of such a concept comes into question, as we should expect it to, it may be for one of two reasons: either because it is an "essentially-contested" concept or because it is a radically confused concept.[9] To invoke the notion of family resemblance must therefore imply that one is prepared to argue for those resemblances.

Yet another question can be raised about the ways in which Nyíri has exploited intellectual historians' analyses of *political* conservatism for the sake of extending it into the cultural and epistemological sphere. This is a note-worthy departure from the procedure that Klemens von Klemperer (one of Nyíri's heroes) followed in writing *Germany's New Conservatism*; it incorporates a move which Klemperer has explicitly disavowed in another of Nyíri's heroes. "Attempts (for instance by Armin Mohler) to broaden the definition of the 'conservative revolution'," he writes, "in order to include such writers as Dostoievsky, T. E. Lawrence and D. H. Lawrence have only helped to obscure the issue".[10] I doubt, therefore, that relating Wittgenstein to this movement will improve the situation. Indeed, Klemperer's service has been to insist that liberal and conservative are correlative *political* notions, which are only meaningful along a political spectrum extending from the radical left to the reactionary right. This has important implications for Nyíri's application of the term. So, it will do well to examine what Klemperer has in mind when he makes this suggestion. Though he does not spell out what his spectrum might look like in detail, I think we can safely conclude that it might resemble the following:

(1) Radical — advocate of total socio-political transformation in accordance with some abstract ideal, i.e. equality.

(2) Liberal — advocate of change with regard to the status quo, which is neither total nor respectful of traditional modes of action.

(3) Conservative — advocacy of those changes which traditions permit, limited opposition to change.

(4) Reactionary — total opposition to change (except resortation of the *ancien régime*).

It is noteworthy that these four categories of individuals define themselves to a significant extent by that which they would *oppose*. This explains why Nyíri can say with justification, "A conservative political creed as such does not exist. Conservative politics change with the times, many of today's conservative aims correspond to liberal ideas of yesterday".[12] To be conservative is always to be conservative in a context and that is to trust traditions and suspect innovations. However, this is to hold views which are in no small part theoretical and abstract seriously enough to oppose it effectively without committing oneself to some theoretical beliefs. We must also bear in mind MacIntyre's point that radical innovations can be espoused in the name of tradition, that to call a measure conservative is not necessarily for it to be conservative. Thus, it is necessary to distinguish between theories which are actually conservative, i.e. do in fact support the existing order or its recognized legitimate extension, regardless of the intentions of their advocates, and those which actually contribute to radical change, i.e. changes in the criteria of legitimacy. In short, we must raise the problem of theory and practice.

The aspects of this problem which most interest me are glanced over by Nyíri without reflection in his efforts to define conservatism without reference to a theory or picture of society. We have seen Nyíri defining conservative theory "depending upon what remains — if anything at all — that is regarded as worth conserving, or even re-establishing".[13] It would seem that the assumption that he makes in this statement would construe conservatism so widely as to make it a vacuous, if not "radically confused", term. This is because it would obliterate the distinctions along Klemperer's spectrum between conservatism and *both* liberalism and reaction. We could no longer distinguish conservatism from liberalism because both are committed to the current criteria of legitimacy *in some loose sense*. It is noteworthy that there is little discussion of the specific views to which conservatism and neo-conservatism are *opposed* in Nyíri's presentation. The distinction between conservatism and reaction would break down on Nyíri's view because in the post-French Revolution era, the world in which we live, reaction must be committed *against* the status quo in a way that conservatism need not be. In short, it makes all the difference in the world, again *pace* Nyíri, whether we believe that the status quo is worth defending, or whether another order is worth re-establishing, for the latter commits us to a reified, abstract notion of society, while the former does not.

Let me try to approach the difficulties I find in Nyíri's conception of the relation of theory and practice from another angle. The question we must ask him is, must a conservative produce a conservative theory (idea, intellectual technique)? I think the answer is *no*. My reasoning is based upon the contingent relationship between the context in which a theory (idea, intellectual technique) is *discovered* and the manner in which it *acquires significance for a community*. The claim that a theory, say, can be described as "conservative", on Nyíri's account, can only be made on the basis of the *intentions* of its discoverer and *perhaps* its implications for society, since he has rejected the possibility of identifying the political (social, cultural) significance of a theory on the basis of its substance. An examination of some parallel cases will reveal that the actual implications of a philosopher's work have, if anything, more to do with the context into which they are received than they do with their originator's intentions. The most dramatic case in point in the last century has been the case of Nietzsche. For Crane Brinton he is a proto-Fascist, if not an out-right Nazi thinker.[14] For Walter Kaufmann he is the antithesis of everything the Nazis stood for.[15] For J. P. Stern fatal confusions surrounding Nietzschean "authenticity" make it possible to derive *both* from his works.[16] During the last hundred years only the Bible has been cited more often in disparate socio-political contexts. It is hardly accidental that the main attraction for both political and non-political, rightist and leftist, rational and irrational

Nietzscheans alike is the perception that the core of his writing is *opposition.* For leftists it is opposition to Christian theology (though, as William McGrath has shown in his brilliant study *Dionysian Art and Populist Politics in Austria* not necessarily opposition to Christianity)[17], for rightists opposition to scientism and socialism and, above all for youth generally, opposition to an established older generation, such as that which informed Moritz Schlick's Nietzschean positivism. The one thing that emerges from this catalogue is that it is virtually impossible to identify Nietzsche's work as conservative, or indeed with any political category, in any unequivocal way, dispite the enormous influence he exerted in conservative circles of all sorts. I suggest that there is hardly a more dubious move than to extend these political metaphors to describe epistemological positions. We shall see this by looking briefly at the self-consciously "anarchist" epistemology of Paul Feyerabend, which, I assume, would have to be characterized as radical on Nyíri's view.

Feyerabend claims that all existing philosophies of science (i.e. Kuhn, Lakatos, et al.) are unduly restrictive and that the best way to respond to the problems to which philosophers of science have addresses themselves regarding the demarcation of the scientific from the non-scientific is to adopt a scientific hedonism and refuse to discuss such problems.[18] We ought, as philosophers of science, to urge recognition of any activity which proclaims itself to be scientific as scientific with only the stipulation that the proponents of science defend their views publicly *á outrance.* For Feyerabend this view, "anything goes" taken to be anti-ideological, progressive and humanitarian. By terming himself an epistemological "anarchist" Feyerabend invites us to locate him in a tradition which begins with the radicalism of the Levellers, continues through the French Revolution (where the term applies derogatorily to anyone farther to the left than the speaker), Godwin and Bakunin to the Spanish radicals in the Civil War. At the same time as he associates himself with this tradition of criticism of the status quo and this vision of the future, however, he hopelessly confuses the matter by asserting that the position he maintains is actually just what we are in fact doing today anyway. If he is not being ironic, it is hard to understand how he is not contradicting himself. If science were merely a matter of internal criticism, the ultimate validity of which could only be determined at the end of a debate, his would be a desirable position. However, the fact that not a little research is done with public funding calls the feasibility of his suggestion into question. From that perspective his suggestion would seem absurd or absurdly utopian and in no sense contemporary practice. He seems to suggest that we should either finance anyone with a research project, which is financially impossible, or modify society in some way such that it will not be impossible to do so. Thus, given the state of society it is hard, for me at least, not to read highly conservative implication in Feyerabend's self-proclaimed "anarchist" epistemology. Would it really be such a radical change to reject the notion that there can be fruitful methodological discussions? In reality does not this entail abandoning science to a mandarin caste even more so than it already is? Is not at least part of the function of methodological discussion of science in an open society to assist in the general public's participation in making science policy decisions? I think so. Whether I have represented and judged Feyerabend correctly remains to be seen. I only hope that I have been able to cast a doubt on the procedure of classifying epistemological theories in political terms. It seems difficult to avoid agreeing with Klemperer that the extension of the term "conservative" from explicitly political contexts only obscures the issues it claims to illumine. We must now turn to the problems in Nyíri's arguments relating Wittgenstein's personal beliefs to his philosophical views.

When Nyíri argues that Ludwig Wittgenstein was personally deeply conservative, that he read conservative writers and writers like Dostoievsky who were generally admired by conservatives, he is on sure ground, for there is little doubt that Ludwig Wittgenstein was tempermentally conservative. He did not regard many of the developments that he experienced in his lifetime as in any sense progressive. Both Fascism and Socialism were equally "foreign and uncongenial"[19] to him, "I often reflect", he wrote in 1929, "Whether my cultural ideal is a modern one, i.e. contemporary, or from the era of Schumann. At the very least it seems a continuation of the latter".[20] However, there is a problem both in drawing inferences in the way Nyíri does from the material he cites — Wittgenstein "wonders" about the Biedermeier but does not wholly identify with it — and also with Wittgenstein's personal individualism and iconoclasm, which would seem to make Wittgenstein, at the very least, a most peculiar type of conservative. First, however, it is necessary to discuss Nyíri's treatment of Wittgenstein as a reader of Dostoievsky. In reminding us of Wittgenstein's profound respect for Dostoievsky, Nyíri also reminds us that the German edition of the great

Russian novelist's works was done by two eminent neoconservatives, Arthur Moeller van den Bruck and Dimitri Merzekowski, Nyíri indicates interesting parallels between Wittgenstein's views and those of Moeller and Merzekowski, which suggest that Wittgenstein *might conceivably* have read their works. Nyíri does the same with Paul Ernst, whom Wittgenstein lamented not having acknowledged in the Preface to the *Tractatus* as the source of the expression, "misunderstanding the logic of language", with a view to strengthening the *plausibility* of his case. Certainly, this sort of thing is important for establishing and understanding Wittgenstein's intellectual milieu and, as such, I have nothing against it. Indeed, I probably qualify better than most philosophers and historians of ideas as a past master of the technique of mustering circumstantial evidence for the sake of enhancing the credibility of a point. However, we have to bear in mind that in Wittgenstein's case there are difficulties in inferring from what we know he definitely read and liked to what he *might* have read and liked. We know, for example, that he read and admired Karl Kraus but considered Albert Ehrenstein's "Krausian" novel *Tubutsch*, in a word "muck".[21] We also know that at one point he admired *Der Brenner* for its public defense of Kraus but, paradoxically, looked down upon its two chief contributors, Theodor Haecker and Carl Dallago, as being among the many who babbled about what could only properly be passed over in silence.[22] So, given Wittgenstein's way of looking at these matters, it is questionable whether he would be impressed with someone merely on the basis of a common interest in, say, a certain writer. Consequently, it should be clear that there is a risk in approaching Wittgenstein as Nyíri does. To be sure, Nyíri is subtle enough to avoid any outright claims which cannot be documented and, indeed, the parallels he presents us with are often both trenchant and enlightening; but, at the same time, he writes in such a way as to invite us to draw the inference that Wittgenstein was in some sense "influenced" by these thinkers.

While this is a relatively trivial criticism, the question of Wittgenstein's absolute ethical individualism raises a more serious problem for Nyíri's thesis. It should be noted that the problem rises from Nyíri's commitment to the notion that there is such a thing as a "conservative attitude", which involves the implicit assumption that the alleged conservative's beliefs are internally consistent. One of the cornerstones of conservatism generally has been a commitment to the notion that moral responsibility is a public matter. This is something Wittgenstein steadfastly denied. He refused to make a moral judgement upon Brutus for slaying Caesar on the grounds that we are not privy to his inmost thoughts and intentions.[23] Whether or not this belief can be squared with his later philosophical position, there is plenty of evidence to indicate that it was among his most cherised beliefs. Further, his general non-conformist attitudes with regard to matter of social convention surrounding dress and the rituals of High Table at Cambridge seem to bear out a deep personal commitment to individualism. His disdain for High Table, incidentally, seems to present us with strong evidence that he had little tolerance for ceremoniousness and seems to present us with strong evidence for resisting Nyíri's efforts to associate him with Adolf Grabowsky's version of conservatism. I must emphasize that I am not claiming that these aspects of Wittgenstein's beliefs and behavior "refute" Nyíri's allegations but merely that it is more difficult to pin that label on him than Nyíri seems to think. Thus, I would never say that Nyíri cannot contend with my objections but that he has not made his case until he has done so.

Nyíri's efforts to set the scene for the development of Wittgenstein's later philosophy is surrounded by similar difficulties. Nyíri suggests that the radical shift in Wittgenstein's thinking "trivially" took place in 1929 but may well have happened as early as his 1925 trip to England. Here again, Nyíri has given us an interesting alternative to the standard way of looking at Wittgenstein's development but he does not come to grips with yet another alternative version of that story, namely, that the transition from the *Tractatus* to the *Philosophical Investigations* was more of a Gestalt shift than a break with his past. Let me spell out what I have in mind here. Maurice Drury makes much of the fact that Wittgenstein once told him that his basic thoughts came to him "early in life", that his famous encounter with Piero Sraffa did not force him to abandon one position for another but to reject one line of thought for another as the core of his philosophizing.[24] In a similar vein K. T. Fann goes so far as to conjecture that the views that Wittgenstein ultimately articulated in the *Philosophical Investigations* may *even antedate* those of the *Tractatus*.[25] To be sure this hypothesis is far from firmly established but it is nonetheless worth considering. It would not be impossible to make a plausible case that a good deal of what became central to the *Investigations* could be the outcome of reflections on the constructivism of Hertz and Boltzmann as well as James's pragmatism, with which Wittgenstein was certainly familiar long before he wrote the

Tractatus. If this is correct, *even* partially, the argument that we ought to look to the Weimar Republic as the setting in which Wittgenstein's later thought develops must be radically emended.

Yet, there is more to be said by way of criticizing Nyíri's view of the context in which Wittgenstein produced his later philosophy. I have deliberately mentioned Nyíri's citing sources from the Weimar Republic as context for Wittgenstein's development because Nyíri seems to assume that this what is true of the Weimar Republic is also true of the First Austrian Republic. However, one of Nyíri's own authorities, Klemens von Klemperer, argues in a brilliant book, which is indispensible of understanding the intellectual history of the First Republic, called *Ignaz Seipel: Christian Statesman in a Time of Crisis*, that Germany and Austria experienced the twenties as differently as they did most of the rest of their modern history.[26] Roughly speaking, the roles of the left and the right were *reversed* from what they were in the Weimar Republic, with the left actually obstructing the implementation of the basic rule of democracy, one man-one vote. In Klemperer's earlier work, cited by Nyíri, he argued that the stimulus for the German neo-conservative movement was primarily the shock that the Revolution of 1918 and its aftermath presented to the intellectuals of the right. While no one would argue that these events were without reverbation in Austria, the "land without a name" had her own distinctive problems during this period as well. So, it is necessary to reiterate the slogan to which all historians of what we commonly term Austria begin their studies, "Austria is not Germany". I am suggesting that Nyíri has to make a case for assuming that what is true of German intellectuals will also be true of Austrians, if he is to draw upon this material to elucidate the context in which Wittgenstein's later thinking developed. Finally, as an Austrian, and especially as a "Krausian" (whatever sense we wish to attatch to the term in Wittgenstein's case) Wittgenstein would hardly have been shocked by the chaos of the post-war world in the way that more naïve Germans might have been. Even Stefan Zweig reports that there was a sense that the Monarchy could not outlast the death of Franz Joseph long before the war.[27] Kraus, of course, never ceased railing against the doom that Austrian corruption would necessarily wreck on society long before World War I. Certainly, I am not claiming that Kraus or anyone else foresaw the state of post-war society as it would actually be but that these considerations make Nyíri's allegation that Wittgenstein's conservatism was "fostered, first and foremost, by his experiences of the post-war period — experiences of a world order that had vanished and of deepest homelessness".[28] Surely, Wittgenstein did experience a profound sense of homelessness but we can perceive this in letters to Russell from Norway already in *1914*.[29] As to the question of the experience of a "vanished world order", I doubt that this could mean anything in Wittgenstein's case, which is in this respect similar to that of Kraus and Loos, beyond admiration for the long lost Biedermeier simplicity.

Each of the considerations, taken separately, is small, possibly even trivial, but taken together they serve to remind us that Nyíri's portrait of Wittgenstein may require significant revision. In order to complete my analysis of Nyíri's argument I must now turn to the philosophical core of his position, the allegation that the notions of rule-following and forms of life, which are at the core of Wittgenstein's later philosophy are essentially conservative. I wish to bring three kinds of considerations to this discussion, data about Wittgenstein's relationship to his philosophizing, textual points about the *Philosophical Investigations* and conceptual points concerning forms of life and rule-following.

While Nyíri rightly draws our attention to the *Kulturpessimismus* expressed in the unpublished Forword to the *Philosophische Bemerkungen*, he fails to see the significance of the very fact that Wittgenstein seems not to have deemed this appropriate to publish. Indeed, it appears to be a characteristic of his works as a whole that he wished to separate his philosophizing from his personal beliefs. We have in recent years discovered more and more evidence that Wittgenstein was personally profoundly conservative and that this conservatism certainly casts new light on his work but I think it also makes it all the more amazing that the only work in which he deliberately and explicitly emphasizes his conservatism, *Vermischte Bemerkungen*, is much more a personal statement which intends to reveal the man's thought's rather than the philosopher's positions. Nyíri emphasizes that the evidence we have to go on is scant but I think that what we do have to go on indicates that Wittgenstein took great pains to separate his cultural pessimism from, say, his view of meaning or thinking. One instance of this occurs in the "Lectures on Aesthetics". There he incites his students to perform the thought-experiment of imaging what happens when a high culture such as German musical culture in the nineteenth century *declines*. Shortly thereafter, Rush Rhees asked Wittgenstein about his "theory of degeneration". Wittgenstein responded with a flat rejection that he maintained such a theory. He went as far as to assert,

"I might approve deterioration — 'all very well your musical culture, I'm glad children don't learn harmony now'".[30] So, there is at least some specific evidence that Wittgenstein did not want to run his personal beliefs together with his philosophizing. I am suggesting, then, that it is not accidental that there are precious few explicit cultural-pessimist remarks in the works which he prepared for *philosophical* audiences. I am inclined to attribute this to his peculiar brand of individualism. The only knowledge he deemed worth having is self-knowledge, so his role as teacher was anything but to convince his students that his opinions were the correct ones but rather, like Socrates Montaigne and Lichtenberg, to stimulate them to think for themselves. In this situation the best strategy is to conceal one's own views.

My second problem with the identification of forms of life and rule-following as conservative philosophically is that there is very little textual evidence that Wittgenstein is committed to an irrationalist account of rule-following as Nyíri maintains. Certainly, the text, "I obey the rule *blindly*,"[31] presents a problem for the position I take but it is not an insurmountable one: However, it is incumbent upon me to offer a "non-conservative" reading of that text. Two things are interesting about the text from that perspective. First, it makes a point that Wittgenstein does not explicitly emphasize for it only occurs once in the *Investigations*. It can, of course, be used as a key to interpreting other passages but we have to justify such readings. The second point is that Wittgenstein is actually emphasizing the contrast between rule-following as a matter of social convention and its misinterpretation as a matter of the *capricious choices of individuals*. In this sense to assert that rule-following is blind is merely to say that it is an orderly procedure in which we must adhere to the prescribed order which the concept designates or we cannot be said to possess the concept. It is not, as Nyíri argues, that the rule cannot be explained or justified *simpliciter* but that it cannot be explained or justified in the sense of being given a "theoretical foundation". There would be no point at all in writing the *Philosophical Investigations* if Wittgenstein did not believe that he in some sense could *explain*, i.e. make intelligible, what it is to follow a rule by convincing us to reflect upon our own behavior, to come to see what it is we *do* when we think. To the extent that we can "see" or "show ourselves" *our own practices and procedures* we can indeed explain our own rule following, i.e. inasmuch as coming to understand it is a way of explaining it.

Another difficulty with Nyíri's argument stems from the fact that he reads more into the notion of "forms of life" than one can find there without doing violence to the text. The very fact that Hannah Pitkin has been able to argue that Wittgenstein's teaching about "forms of life" can be constructed as the indispensible prerequisite to the kind of understanding of society that alone makes radical social change a genuine possibility is *prima facie* evidence that Nyíri is reading an interpretation *into* the text.[32] However, before elaborating on that point we ought to look at what Nyíri actually says about forms of life. "The decisive point", he argues, "is that we cannot entertain a liberal attitude as regards irregularities in our *own* society. For it is through compelling uniformities that the life of a society becomes *ordered*; such uniformities determine the *boundaries* of society — i.e. only through such uniformities does the society as such become constituted".[33] There is a subtle shift from Wittgenstein to Nyíri's reading of him Nyíri identifies "forms of life" with "societies". Actually, Wittgenstein tells us precious little about these "forms of life".

Briefly, there is no guarantee that what Wittgenstein means by form of life is anything other than a set of social practices. Therefore, it is *contingent* whether our normal uses of "culture" or "society" correspond to Wittgenstein's forms of life. It must be remembered that Wittgenstein nowhere gives us a nuanced account of just what constitutes a form of life nor does he provide us with any extant examples. To do so was not part of the problem to which he was addressing himself. It is necessary to remind ourselves to this from time to time. Forms of life enter his *Investigations* as a mode of explicating the contextual, praxis-orientation of *meaning*. Wittgenstein's achievement in that work was to have supplanted Russell's (and his own) syntactic-formalist account of meaning by confronting it with an explication of concepts in terms of their concrete cultural ecology. Actually, Wittgenstein may not even have had anything like a clear cut concept of just what "forms of life" are after all. This may seem strange but it is by no means inconsistent with his personal psychological problem-solving apparatus. Recall that his achievements have been largely the result of his extraordinary ability to formulate and reformulate alternatives to the views he criticizes by this relentless pursuit highly *specific* points of criticism, i.e. from his amazing ability to focus his mind's eye *very narrowly*. If the results of this ability to circumscribe his vision have astonished us, we should not be astonished to discover that the scope of his extraordinary vision should be in some respects narrower than our own. We should not be surprized, then, when things that would seem to us as part and parcel of his approach to meaning should not have occurred to him. We should bear in mind that our problems often arise from his *solutions*. However, there is actual evidence that the sort of lack of mental peripheral vision I have been alleging to have been part of his personal psychological constitution affected his approach to another problem. Wittgenstein himself was astonished to realize that he had not been able to give any examples of the kinds of things he discussed under the rubric of "objects" in the *Tractatus*. His excuse for not being able to do so was simply that at that time he was thinking as a *logician*, and as such this simply was not his concern.[34] This is amazing in the light of the fact that the problem of objects has been one of the most criticized aspects of the *Tractatus*. Thus, I see no reason for thinking that, were he alive today and asked why he did not identify and discuss existing forms of life in the *Investigations*, the answer would be any different from his response to the query about the objects of the *Tractatus*. The fact of the matter is that though Wittgenstein tells us how forms of life function as constituting meaning and how to go about investigating them, i.e. by identifying expression within language games and then determining these latter within their appropriate forms of life, he never does so. Nor, I would maintain, should he be expected to. It simply was not his problem.

To return to my point, there is no reason to believe that the forms of life we identify on the basis of our "semantic anthropology" will be identical with what we conventionally or historically term cultures or societies. Indeed, a good part of the history of Central Europe since the French Revolution has been a history of the revolt of "semantic communities", constituted by agreement in judgment as Wittgenstein calls them against political (and economic?) society. Be that as it may. If we want to do something other than what Wittgenstein does with the notion of forms of life we have to begin with the sociological and anthropological suggestion of such thinkers as Basil Bernstein and Malcolm Crick (from whom I have borrowed the term "semantic anthropology").[35] Their work begins with the notion that we really know very little about the cultures and subcultures which are constituted by shared linguistic expression. Not a little of my skepticism about identifying form of life and the ordinary usage of "society" come from Bernstein's insistance that I both do and do not share meanings with, say, speakers of "Black English" who also happen to reside in the Commonwealth of Massachusetts. A further point, there is no guarantee that all forms of life are the same except inasmuch as each forms a context to which individual expressions must be referred if they are to be grasped. There is no reason to believe that they would be similar in other respects, contrary to what Nyíri suggests.

It follows from the fact that we have not as yet actually identified any concrete forms of life that we cannot possibly be in a position to assess Nyíri's first claim that the given form of life cannot be transcended, his second claim that the possibility of other orders does not in the least weaken the inexorable binding force of our own, and finally, whether or not all changes in forms of life are autonomous developments. Because these were not Wittgenstein's questions, we should not expect anything apart from *hints* in his works as to how we ought to go about making judgments in these matters. Surely, Wittgenstein believes that forms of life change because we know there have been drastic changes in individual expressions' meanings. However, it is a matter of having a look and describing these changes (if we can understand them) whether or not Nyíri's thesis is tenable. It does not seen *inconceivable* that there are forms of life which learn *something* about other forms of life and make adjustments in their own.

Hannah Pitkin has argued eloquently that it is anything but conservative to maintain that the indispensible prerequisite for any radical social change is an accurate assessment of existing conditions. She maintains that this will be to look to the ways in which we are *constrained* by the existing order — including conceptually. We must look, she suggests, to the learning of speech acts and, specifically, political speech acts to see the difference between what is really "conservative" philosophically and politically. She sees no reason for failing to identify the political and a-fortiori cultural implications of Wittgenstein's *Investigations* with the view expressed in the following text and I see no reason to disagree with her: "Men make their own history, but they do not make it just as they please. They do not make it under circumstances chosen by themselves, but under circumstances directly encountered, given and transmitted from the past". I am prepared to assent to Nyíri's designation of Wittgenstein as conservative to the extent that he will extend the term to cover the author of these phrases, Karl Marx.[36]

NOTES

[1] G. H. von Wright, "Wittgenstein in Relation to His Times", *Wittgenstein and His Times*, ed. B. F. McGuinness (Chicago, 1982), pp. 108—20.
[2] J. C. Nyíri, "Wittgenstein's Later Work in Relation to Conservatism", *Wittgenstein and His Times*, p. 57.
[3] *Ibid.* p. 44.
[4] *Ibid.* pp. 45—6
[5] Alasdair MacIntyre, *A Short History of Ethics* (New York, 1968), p. 227.
[6] E. P. Thompson, *The Making of the English Working Class* (Harmondsworth, 1968), p. 603 *et passim.*
[7] Sigmund Neumann, "Forward", Klemens von Klemperer, *Germany's New Conservatism* (Princeton, N.J. 1957), XII.
[8] Clinton Rossiter, "Conservatism", *International Encyclopedia of the Social Sciences*, Vol. **III**, 290.
[9] W. B. Gallie, "Essentially Contested Concepts", *The Importance of Language*, ed. Max Black (Engelwood Cliffs, 1962), pp. 121—46.
[10] Klemens von Klemperer, "Conservatism", *Marxism, Communism, and Western Society*, Vol. **II**, 1967.
[11] Klemperer, *Germany's New Conservatism*, p. 5.
[12] Nyíri, *op. cit.*, p. 46
[13] Nyíri, *loc. cit.*
[14] Crane Brinton, *Nietzsche* (Cambridge, Ma., 1941).
[15] Walter Kaufmann, *Nietzsche* (Princeton, N.J., 1950).
[16] J. P. Stern, *Hitler: The Führer and the People* (Berkeley and Los Angeles, 1975), pp. 44—9 *et passim.*
[17] William McGrath, *Dionysian Art and Populist Politics in Austria* (New Haven, 1971).
[18] Paul Feyerabend, *Against Method: Outline of an Anarchist Theory of Knowlede* (London, 1978).
[19] Ludwig Wittgenstein, *Vermischte Bemerkungen*, ed. G. H. von Wright (Frankfurt am Main, 1977), p. 20.
[20] Wittgenstein, *ibid.* 14
[21] Wittgenstein, *Letters to Paul Engelmann with a Memoir*, ed. B. F. McGuinness, trans. L. Furtmüller (Oxford, 1967), p. 4.
[22] Wittgenstein, *Briefe an Ludwig Von Ficker*, ed. G. H. von Wright (Salzburg 1969), pp. 32—8.
[23] Rush Rhees, "Some Developments of Wittgenstein's View of Ethics", *Philosophical Review* **74** (1965), 22.
[24] Maurice Drury, *The Danger of Words* (London, 1973), IX.
[25] K. T. Fann, *Wittgenstein's Conception of Philosophy* (Berkeley and Los Angeles, 1969), p. 42.
[26] Klemperer, Ignaz Seipel: *Christian Statesman in a Time of Crisis* (Princeton, N.J. 1972) pp. 209—10 *et passim.*
[27] Stefan Zweig, *The World of Yesterday*, trans. anon. (New York, 1943), p. viii.
[28] Nyíri, *op. cit.* p. 49

[29] Wittgenstein, *Letters to Russell, Keynes and Moore*, ed. G. H. von Wright (Ithaca, N.Y. 1974), p. 47.

[30] Wittgenstein, *Lectures and Conversations on Ethics Psychology and Religious Belief*, ed. Cyril Barrett (Berkeley and Los Angeles, 1967), p. 10.

[31] Wittgenstein, *Philosophical Investigations*, trans. G. E. M. Anscombe (Oxford, 1967), Part I, par. 219.

[32] Hanna Pitkin, *Wittgenstein and Justice* (Berkeley and Los Angeles, 1972), pp. 338—39.

[33] Nyíri, *op. cit.*, pp. 60—1

[34] Fann, *op. cit.* p. 42

[35] Basil Bernstein, *Theoretical Studies Towards a Sociology of Language* (3 vols.: London, 1970); Malcolm Crick, *Explorations in Language and Meaning* (London, 1976).

[36] Karl Marx, "The Eighteenth Brumaire of Lous Bonaparte", *The Marx-Engels Reader*, ed. Robert C. Tucker (New York, 1978), p. 595.

CHAPTER III

WITTGENSTEIN, MARX AND SOCIOLOGY

The very act of mentioning Wittgenstein's name in connection with either Marx or sociology, let alone both simultaneously, is apt to disturb any self-respecting Wittgensteinian. On the face of it nothing could seem farther from Marx's celebrated exhortation to philosophers to change the world than Wittgenstein's admonition that "philosophy may in no way interfere with the use of language . . . it leaves everything as it is".[1] Furthermore, what could form a sharper contrast with Marx's allegation that history is the history of class *struggle* than Wittgenstein's claim that it would be pointless to advance theses in philosophy because everybody would agree about them.[2] Wittgenstein's personal conservatism and his abiding concern with the works of such conservative authors as Spengler, Dostoievsky, Tolstoi and Weininger has led J. C. Nyíri to assert that Wittgenstein's mature thought is nothing less than a defense of a peculiar brand of neo-conservatism.[3] Whatever we may discover about Wittgenstein in the future, it is most unlikely that we shall ever turn up the slightest interest in politics let alone political activism. Finally as regards sociology, early in his career Wittgenstein maintained that "the word 'philosophy' must mean something whose place is above or below the natural sciences, not beside them".[4] Anthony Kenny has maintained that this was one of many issues that Wittgenstein never changed his mind about.[5] There is little reason to think that Wittgenstein would see philosophy as any closer to the social sciences. This is because the sciences seek to establish the way things are in the world in terms of propositions which are true; whereas philosophy aims at understanding how this is possible. Philosophy aims not at truth but at clarity, what Wittgenstein later called "perspicuous representation", of the conditions which make truth possible (i.e., our practices).

However, to have said all of this is only to have established that we should not be too hasty to see connections between Wittgenstein and Marx and between his philosophizing and social science; rather, it is only to have suggested that these connections are not easily established. In short, there are responses to each of these objections to associating Wittgenstein with Marx and sociology.

First of all, Wittgenstein was anything but a revolutionary; yet, for all that, he was no less desirous than Marx to see the end of bourgeois society. Indeed, he once remarked that he would rather see a change in the way men lived, which made the sort of philosophizing he engaged in superfluous than a continuation of his work.[6] Moreover, in a remark concerning Frank Ramsey's reluctance to challenge the assumptions of contemporary mathemathics Wittgenstein suggested that this refusal to 'go to the root' was nothing less than reprehensibly bourgeois.[7] It must be hastily added that Wittgenstein appears to have been convinced, unlike Marx, that neither he nor anyone else has the slightest idea of why societies develop in the ways they do. It is no less important, on the other hand, to point out that for Marx himself philosophy was by no means capable of changing the world on its own. Philosophy *and* the proletariat, *Capital* and the International, were necessary for that.

In the second place, there is more than a little evidence that Wittgenstein was personally acquainted with a number of avid, intelligent Marxists, some of whom exerted a powerful influence upon him. For example, Fania Pascal (whose husband, Roy, was himself something of a Marxist) writes of Wittgenstein's deep respect for and friendship with the "fiery Communist", Nicholas Bachtin.[8] George Thomson, whose Marxist interpretation of pre-Socratic philosophy scandalized classical scholars, also numbered among Wittgenstein's friends. Surely, Wittgenstein did not lack the opportunity to become familiar with Marx's views. However, much more significant for making a case for Wittgenstein's familiarity with and a possible influence from Marx is Wittgenstein's acknowledged debt to Piero Sraffa, whose *Production of Commodities by Commodities* remains a classic of Marxist analysis. Of Sraffa Wittgenstein wrote in the Preface to the *Philosophical Investigations* that he was "indebted to *this* stimulus for the most consequential ideas" in his book.[9] On another occasion he described Sraffa's effect upon him as having made him feel like a tree all of whose branches had been cut.[10] Unfortunately, neither Wittgenstein nor Sraffa were inclined to say more about the nature of their interaction.

Clearly, then, it is less than implausible that Wittgenstein's work can be seen in relation to that of Marx. As Ferrucio Rossi-Landi has written, "when we look deeper than the current historiographical divisions permit, there is a sense in which Karl Marx and Ludwig Wittgenstein may even appear complimentary or, in certain aspects, actually at one".[11] If this is the case, it will be of the utmost importance carefully to examine allegations of complementarity in their thinking. Moreover, a variety of thinkers such as Hannah Pitkin, John Danford, Derek Phillips and Alan Ryan have alleged that there is a great deal in the later Wittgenstein that can illuminate a variety of issues in social science. Perhaps the most radical claims to find complimentarity between Wittgenstein and Marx and at the same time illuminate basic *quandaries* in the social sciences have been advanced by David Rubinstein.[12] For that reason it will be particularly useful to examine his arguments carefully. Getting straight about the claims that Rubinstein has advanced will not only cast light upon Wittgenstein's relationship to Marx and to sociology but clarify his relationship to a number of other thinkers, most of the chief problems involved in such a 'use' of Wittgenstein and the all-important issue of Wittgenstein's conception of the relationship between philosophy and theory in Wittgenstein's work. Contrasting his appropriation of Wittgenstein with that of Pitkin will add sharpness to the picture of a Wittgensteinian social philosophy, which will be rounded out with a discussion of the role of the crucial notion of 'forms of life' in that undertaking.

Rubinstein's claim is that Wittgenstein's contextualism and his peculiar emphasis upon social practice taken in conjunction with Marx's historical and economic analysis provide a sounder account of the *subject* that sociology should study, than any of the alternatives currently available. Thus, Wittgenstein's account of meaning in terms of 'forms of life' and his emphasis upon how traditional epistemological problems could be resolved by coming to comprehend social *practices*, instead of theoretically, is taken to complement the Marxian notion of praxis. Understanding our own actions correctly makes the theoretical activity of providing firm foundations for knowledge superfluous according to both Marx and Wittgenstein. But what are the alternatives which have been displaced?

Rubinstein's answer is both unusual and revealing of the lengths one must go to fit Wittgenstein into contemporary social theory. The two displaced alternatives turn out to be approaches to sociological theory whose faults are rooted in defective accounts of *mind.* They are referred to as "objectivism" and "subjectivism". Objectivism is a behaviorist approach to sociology which relies principally upon surveying techniques whose results are correlated according to "covering laws". Subjectivism, on the other hand, takes the chief aim of the social sciences to be the articulation of common beliefs and the interpretation of conscious action. Despite the fact that nearly all of contemporary sociology falls into one or the other classification Rubinstein contends that there are insurmountable obstacles to maintaining either view. The former glosses over its own interpretive biases; while the latter fails to recognize the problem of false conciousness. Where the one produces surveys whose question contain descriptions of situations which more or less predispose those questioned to answer in a particular way; the other fails to recognize that the ways in which we describe our own behavior are not always accurate. The alleged fallacy common to both approaches is the common assumption that mind and matter are two different things. It is not that the objectivist or the subjectivist accepts this dualism openly. Rather it creeps in as each rejects one part of the mind/matter dichotomy to accept the other. Thus, in rejecting the distinction both presuppose it and, consequently, are determined by it without being aware of that fact. Both objectivist and subjectivist remain haunted by the specter of Cartesianism. At this point Rubinstein claims that *only* a synthesis of Marx and Wittgenstein can finally overcome the shortcomings which result in social theories determined by the ghost of Descartes:

> Marx and Wittgenstein reject dualism with the claim that the mind is not a private entity. Rather, they show that mind is a property of the social conduct of real, active and socialized persons. In concert with this view, they show that meaning is not a feature of subjective experience but of systems of social praxis. In contrast to the traditional philosophical doctrines that have played such a central role in social scientific thinking, Marx and Wittgenstein show that the meaning of an action does not lie in subjective experience but in the surrounding social context and that an understanding of action hence presupposes an understanding of the social context in which it occurs.[13]

On this view Marx needs Wittgenstein because, for all his insight into social life and his avowed anti-dualism, Marx left his criticism of Cartesianism at too general a level. Wittgenstein needs Marx because his idea of 'forms of life', for all its concreteness, provides little guidance as to which features of the social environment ought to be seen as decisive for its description. From the synthesis of the two philosophies of praxis — and from it *alone* — can a proper account of social explanation be derived. Thus, a very strong two part claim is being made. First, sociology needs a theory which will do justice to the description of a social milieu in such a way that its members' shared understandings will be thoroughly articulated but which will also be capable of guiding fieldworkers with respect to those features of social behavior to which members of the society under study are themselves blind. Second, only a synthesis of Marx and Wittgenstein can accomplish this task.

There are several reasons for doubting that the central assumptions upon which these claims rest will stand up to examination. Yet, the thesis is sufficiently provocative and important that a critique of the arguments upon which these claims rest can contribute to a deeper understanding of Wittgenstein, Marx and sociology. Let us begin by examining his view of the current state of social theory.

For a start, there can be little doubt that it is insightful to link conceptual problems in social theory to assumptions about the nature of mind. This is nothing else than to realize that the problems in social theory are at least in part *philosophical* ones. In any discipline, apart from models of how to proceed and analyzing just why successful works are successful, there is a necessity to look at how unarticulated universally accepted assumptions, what R. G. Collingwood termed "absolute presuppositions", frustrate the goals the discipline sets for itself. This task is the philosophical component in any theoretical endeavor. It is the task to which Rubinstein has addressed himself.

He offers what he terms a *rough* contrast between the two types of social theories in the following table:[14]

objective	subjective
materialism	idealism
explanation (covering law)	understanding (*Verstehen*)
observation	interpretation
facts	ideas
cause	meaning
behavior	action
existence (production)	consciousness (ideology)
substructure	superstructure
social organization	social psychology
structure	culture

Unsurprizingly, Durkheim is the paradigm case of the objectivist theorist; whereas Weber is the paradigm of the subjectivist. However, Rubinstein assures us that we should not expect to find that anybody corresponds to all of these characteristics. His basic problem is that, although both represent important points of emphasis, both are radically incomplete to the extent that they rule out irreducible elements within social experience. This is where the demand for a new *type* of theory arises, one which would retain what is legitimate and vital in both objectivism and subjectivism and at the same time eliminate the blinkered perspective, which radically limits each. However, in his zeal to establish the conceptual roots of both programs he seems to *caricature* both.

Rubinstein is aware that he will be charged with caricature in his effort to produce a sharp picture. As a result, he makes considerable efforts to establish that the pronouncements which, say, a Durkheim makes about his 'positivist' methods should by no means be construed as representing his actual practices as a sociologist. However, caricature seems to creep in when Rubinstein discusses the Diltheyan conception of empathy. He assumes that this notion is central to any subjectivist sociology. It is crucial to this thesis that empathy be defined very precisely as the attempt to relive the experience of the other. This, he claims, is impossible. Yet, this seems to ignore the practices of those sociologists who take the idea seriously. The practice of reliving another's experience has nearly always, in those cases where it has been fruitfully employed, been one of trying to understand another's *problems, questions, perplexities* and *conundrums*. In those cases where such a method has paid off the results which have been obtained come more from understanding how a grasp of someone's questions leads us to the *criteria* they employ in determining relevance with regard to their ideas or activities. It has not been a matter of subjective feeling at all but an effort to articulate à la Collingwood the assumptions implicit in the questions of those whom we study, a matter which accessible to public scrutiny and not at all something esoteric. For all his rhetoric concerning "private meanings" and the like, Max Weber, for example, by no means ignored either 'objective' social structures nor did he ignore the larger context in which those meanings were embedded. Rubinstein is well aware of this and, consequently, emphasizes (unfortunately without illustrating) the tensions in the works of subjectivist social scientists which parallel those in the work of objectivists. "Like the objectivist, the subjectivists routinely violate their philosophical pronouncements Rubinstein asserts. Yet, he proceeds to defend his distinction, "nonetheless the philosophical roots of subjectivism remain visible in much of the work carried out in this perspective".[15]

This last passage seems to concede that there is at least some degree of caricature in clinging to the subjectivist-objectivist dichotomy. This element of caricature more than anything else casts a shadow over his efforts to link his critique of social theory to philosophy of mind. It also seems to provide a much-needed clue to what he is really up to; namely, bringing the theory and practice of sociology closer together by eliminating residual ideological elements emanating from the positivist claim that there is at bottom one and only one scientific methodology. Without explicitly saying so, Rubinstein is asserting that polarization over this notion has hindered the growth of sociology by introducing a cleavage at the theoretical level, which has only confused, rather than clarified, the task of sociology. In effect, his efforts at wedding Marx and Wittgenstein are based upon a confidence in the anti-ideological influence that a concept of social explanation drawn from a union of their perspectives upon social praxis will have.

Seeing Rubinstein's thesis from this vantage point helps to illuminate the question of why he is so insistant that only a marriage of Wittgenstein and Marx can accomplish the desired de-ideologization. Indeed, one of the interesting features about this treatment of Marx, in contrast to his treatment of Durkheim and Weber, is that: whereas the latter are taken to be inconsistent with respect to matters of theory and practice within sociology, Marx is taken to be utterly unambiguous with respect to his pronouncements and his practices. In short, the tensions in the works of Durkheim and Weber are highlighted while those in the works of Marx — those very ambiguities which make interpreting Marx a cottage industry — are effectively suppressed. It is not for nothing, after all, that the fathers of sociology, almost without exception, took Marx to be an objectivist of the first order, i.e., a mechanistic, materialist, functionalist. If it is an error to view Marx this way, it is one which, as Alasdair MacIntyre has pointed out, began with Friedrich Engels and one which Marx did precious little to correct.[16] Briefly, if Durkheim or Weber was confused about or misrepresented himself in a way that Marx can enlighten, we should by all means use a Marxian perspective to bring the resulting tensions in their work to the surface. However, by parity of reasoning, if there are similar problems in the texts of Marx himself there seems to be no reason to deny the legitimacy of the Weberian or the Durkheimian in using his conceptual approach for bringing them to the surface. If Marx is not immune to the diseases which afflict Durkheim and Weber, it hardly seems that his work is somehow more fundamental than theirs. If this line of reasoning is correct, the picture of Marx as mediating between objective and subjective approaches to sociology is as much a *construct*, i.e., a function of highly selective interpretation, as his Durkheim or his Weber. This will be the case even if we take his view of Marx to be the correct one.

If all of this is correct, a number of pressing questions present themselves. Why Wittgenstein? Could other philosophers have fulfilled the role Rubinstein has chosen for Wittgenstein? To what extent does Rubinstein's thesis agree with the efforts of other thinkers to unite Wittgenstein and Marx or to relate Wittgenstein's work to problems in the social sciences? Given Wittgenstein's emphasis upon the social nature of meaning etc. does his philosophy ultimately merge with sociology as Rubinstein maintains? To what extent was Wittgenstein a linguistic *philosopher*? Finally, can Wittgenstein's notion of 'form of life' bear a Marxian interpretation?

It is necessary to begin with the question of how Rubinstein goes about connecting Wittgenstein and Marx. He proceeds from a criticism of the view of Wittgenstein in *Wittgenstein's Vienna*, where it is alleged that Wittgenstein ought to be seen, not as a British philosopher but as a Continental thinker and in particular as a Kantian. Rubinstein wants to suggest that "Marx, and perhaps Hegel, ought to be seen as important figures in the *historical* (my emphasis) background of Wittgenstein's thought".[17] He goes on to suggest that there are parallels with Dilthey as well. A couple of points are in order here where it is possible to speak magisterially. First of all, identifying Wittgenstein as a Kantian was a matter of looking at the actual background figures who influenced the author of the *Tractatus Logico-Philosophicus* such as Heinrich Hertz and Ludwig Boltzmann, who might account for the Kantian flavor that had long been recognized in its text. It was, thus, an historical and textual point rather than a speculative one. Secondly, the very aim of identifying a Kantian element in the background to the early Wittgenstein was intended to make plausible the more or less Hegelian turn of the later Wittgenstein. Identifying a Kantian element in Wittgenstein, then, was by no means to be construed as something that was done at the expense of Hegel but for the sake of showing how it could be the case that Wittgenstein could exhibit characteristics of that very Hegelian (and a-fortiori Marxist) tradition with which he had otherwise no contact. We are inclined to forget that there is a more or less direct line of philosophical transmission from Kant to Hegel to

Marx, which a philosophical genius of Wittgenstein's magnitude might be able, at least in part, to recapitulate in his own thinking once he had reached the right point of departure. However, be that as it may, the third point about Rubinstein's efforts to make a connection between Wittgenstein and Marx is that no historical evidence exists to attest to Wittgenstein's familiarity with the texts of either Hegel or Marx despite the researches of the Marxist-Wittgensteinian Ferrucio Rossi-Landi. So, it is not that the authors of *Wittgenstein's Vienna* were in any way opposed to the idea that Wittgenstein came close to central notions in Hegel or Marx but that the book's argument took another direction. In fact, the book was a kind of propaedeutic to just the sort of reading of Wittgenstein that Rubinstein endorses.

To proceed to the text issue: Does Rubinstein really need Wittgenstein to establish a social theory which will avoid the pitfalls of both objectivism and subjectivism? Wittgenstein once complained that the only thing that a certain philosopher, who published a book entitled "_____ *and Language* had learned from him was to say "and language".[18] There is a corresponding tendency today to add "and Wittgenstein" to the title of a book. To discern whether it is more fashion or something deeper which moves Rubinstein to unite Marx and Wittgenstein we should ask, "what does Rubinstein want from Wittgenstein?". He answers the question by insisting that the only adequate rejection of dualism requires an account of mind as embodied. In addition an adequate philosophy of mind requires:[19]

(1) a concept of persons as preeminently actors and, hence, a focus upon actions rather than states of consciousness,
(2) a dialectical emphasis on the surrounding context as the necessary ground for understanding persons; and,
(3) The conception of persons as social beings.

However, there seems little doubt that any number of modern philosophers could meet these demands.

Hegel is the first one who comes to mind. Rubinstein certainly gives Hegel his due. However, in doing so he curiously fails to refer to the significant similarities between Wittgenstein's rejection of the idea of a private language and Hegel's rejection of the notion that sense data are perceived without the mediation of language, which have been brilliantly analyzed by Charles Taylor.[20] The Heidegger of *Sein und Zeit* certainly also deserves mention. His whole aim is to lay bare the ontology of action and in doing so he emphasizes both the contextual nature of meaning and the social character of human nature.[21] The "radical empiricism" of William James, as well as his pragmatism, which we know influenced Wittgenstein, not to mention the more radical social pragmatism of John Dewey would also fill the bill nicely. So, there is no lack of philosophical alternatives to Rubinstein's selection of Wittgenstein as the complement to Marx *par excellence*. Indeed, Richard Rorty has emphasized the ways in which Wittgenstein, Heidegger and Dewey can be perceived as dove-tailing to form an odd unity.[22] So we should not be surprized to discover that, given Rubinstein's account of what he wants from philosophy, he might have chosen other figures than Wittgenstein as complimenting Marx.

It could be objected with some justification that there are *prima facie* grounds for taking nearly all of these figures as being on one or another side of subjective/objective dichotomy. The point of selecting Wittgenstein cannot be merely that Wittgenstein is the *only* thinker who fits the requirements, which are basically Marxist, but that he is the only one of these thinkers who unequivocally and *powerfully* rejects both objectivism and subjectivism. Heidegger *looks* too much like a subjectivist; while Dewey *looks* too much like an objectivist. Wittgenstein, on the other hand, is uncompromizingly opposed to both views as over-simplifications *and* no less uncompromizingly anti-ideological. Thus, to wed Wittgenstein and Marx is in a sense to bring to the fore certain methodological questions which lurk side by side in the huge Marxian corpus with his political activism. Although this is not the way Rubinstein describes himself, this is what he in fact accomplishes. There are good reasons, (which will be touched upon below) for a social theorist to want to do so.

However, that Rubinstein is somewhat precipitous that there are connections between Wittgenstein and Marx, which can be usefully exploited by social theory can be seen by looking briefly at two other efforts to look at Wittgenstein together with Marx, those of Hannah Pitkin and Ferruccio Rossi-Landi, as well as J. C. Nyíri's efforts to claim Wittgenstein for Neo-Conservatism. These three writers are hardly unanimous in their respective estimates of Wittgenstein's relationship to Marx.

In Pitkin's interesting, if misnamed, study of the implications of analytic philosophy in the Wittgenstein-Austin-Ryle mode, *Wittgenstein and Justice*, there is an effort to see Wittgenstein as engaged in the same process of advancing the cause of human self-awareness as Marx and

Freud and Nietzsche.[23] Pitkin's work is notable for the caution with which she associates Wittgenstein with Marx. She rightly sees all of the thinkers mentioned as devoting themselves to disabusing us of our desire to see the world as other than as it actually is, i.e., from comparing an uncomfortable reality unfavorably with a picture of how things should be, if they could be the way we want them to be. On her account, Marx and Wittgenstein want to turn our heads from the clouds, to make us see life as it is lived in concrete social circumstances. What makes Wittgenstein special in this effort is his steadfast refusal to offer the slightest glimpse of alternative possibilities. Moreover, the very form in which Wittgenstein expressed himself, she emphasizes, is designed to highlight the difficulty of obtaining a realistic perspective on ourselves. Not only has Wittgenstein, unlike Marx, no program or vision of the future, but his very teaching is deliberately expressed in a form which is only accessible to a few serious students. Wittgenstein, in short, forces us to be serious about the rootedness of our existence in 'forms of life'. The implication is that the alternative vision of Marx overwhelmed this equally important concern that we confront the world as it is and that we need Wittgenstein to remind us, first, that "not every change is possible" and, second, that "philosophizing will not change our concepts, habits or institutions".[24] Briefly, she cautiously suggests that a Wittgensteinian perspective may be a necessary prerequisite for the sort of radical transformation of society that Marx envisioned. However, in attempting a sketch of what a Wittgensteinian political philosophy would look like she makes an observation which is crucial for the Rubinstein thesis; namely, that a Wittgensteinian political philosophy would have to be written from a perspective *within* society.[25] She rightly takes this to distinguish Wittgenstein from Marx, who, for all his radical innovations in social theory, remains more or less in the role of an observer, i.e., outside of what he describes. This certainly presents a difficulty for uniting Wittgenstein and Marx in the way that Rubinstein would. Even though both are philosophers of praxis, Marx writes in a voice sufficiently similar to that of classical political theorists that the distinction between his approach and that of, say, Hobbes or Machiavelli is difficult to discern stylistically. With Wittgenstein there is never a question that what is sought at the end of his investigation is something radically different from what classical philosophers aimed at.

Ferruccio Rossi-Landi's "Towards a Marxist Use of Wittgenstein", for all its fragmentary character, remains one of the most stimulating discussions of Wittgenstein's relationship to Marxism.[26] Rossi-Landi's avowed purpose is to appropriate Wittgenstein's later philosophy for Marxism. He asserts that Wittgenstein's notion that traditional philosophical language as 'language gone on holiday' ought to be construed as an analysis of "linguistic alienation". Rossi-Landi, who goes unmentioned in Rubinstein's study, writes: "Marx gives the indispensable theoretical framework, Wittgenstein particular elaborations of it as applied to language".[27] Rossi-Landi develops a picture of Wittgenstein as a kind of Neo-Feuerbachian, whose insights can be appropriated by Marxism provided that they are subjected to a Marxist critique. In the end, Rossi-Landi's Marxist critique of Wittgenstein brings the reunion of Wittgenstein and Marx in social theory into question. He rightly emphasizes the parallels between Wittgenstein's notion of language as embodying rationality and Marx's view that language is "practical consciousness" in the German Ideology[28] (a work which was translated into English by Roy Pascal, husband of Wittgenstein's Russian teacher). However, Rossi-Landi insists that Wittgenstein's account of linguistic practices puts language in a public context without noticing that this public dimension is socio-political. This point is surely well-taken. It has to be seen as presenting an obstacle to the marriage of Wittgenstein and Marx. At the very least it seems to imply that a Wittgensteinian Marx would be a radically depoliticized Marx. In any case, Rossi-Landi's Marxist enthusiasm for Wittgenstein in no way emphasizes the possibility of creating a new kind of social theory out of a synthesis of the two. In fact, Rossi-Landi argues against seeing them on the same level as Rubinstein would.

While J. C. Nyíri's efforts to link Wittgenstein to Neo-Conservatism do not directly contradict Rubinstein's thesis, Nyíri's view of Wittgenstein's philosophizing as rooted in his all-pervasive desire to show us how we are encapsulated in our 'forms of life' through blindly following rules that we can never hope to alter, tenuous as it is, certainly adds to our doubts about the legitimacy of Rubinstein's project. At the very least Nyíri's erudite arguments ought to remind us that any efforts to appropriate Wittgenstein on the part of Marxists will have to be argued for in considerable detail. Above all, it will be necessary to show how Wittgenstein's allegation that we follow rules blindly[29] is compatible with any sort of 'scientific' understanding of our own society. This is by no means a devastating criticism of Rubinstein but it ought to serve to remind us of the surface distance which separates the two thinkers. However, we shall only be in a position to see whether a 'deep' level of compatibility between the two exists when we have established just what a Wittgensteinian social philosophy would look like.

The question of Wittgenstein's relationship to sociology is no less problematic than that of his relationship to Marx. In the course of developing his position on this issue Rubinstein warns us that we should not be too hasty to follow common usage in designating Wittgenstein as a 'linguistic philosopher' in the sense that, say, John Austin was. This means in the sense that he was a philosopher who saw the problems of philosophy as strictly linguistic in character in the sense that overtly philological arguments can be seen to resolve philosophical difficulties. Rubinstein fears that this conception of Wittgenstein's philosophizing will obscure Wittgenstein's emphasis upon the rootedness of language in social practice and, consequently, the essentially sociological character of his philosophy. The half-truth of this allegation is sufficiently interesting and important to require discussion.

To be sure, the surface distance between Wittgenstein and Austin seems so great that Stephen Toulmin could write: "The transition from Wittgenstein's Cambridge classes and 'at homes', to the style of philosophical disputations then current at Oxford was a shock. For all its ingenuity and discrimination, the 'linguistic analysis' of Oxford in the 1950s seemed — by comparison with Wittgenstein — to lack any philosophical main-spring".[30] Moreover, John Passmore has pointed out in his magisterial history of English philosophy that the differences between Austin and Ryle — let alone Austin and Wittgenstein — are sufficiently great to make any catch-all term like 'linguistic philosopher' suspect.[31] At the same time this point should not be overstated; for the very same Stephen Toulmin's subsequent philosophizing has to a great extent involved synthesizing Wittgenstein's concern with rule-following with Austin's concern for linguistic nicety and nuance.[32]

Indeed, Wittgenstein's "Remarks on Frazer's *Golden Bough*" are written from a perspective which is close to Austin's insistance that, in understanding an action, the adverbial modifiers, which tell us *how* something is done, are absolutely crucial.[33] In short, there is a complimentarity in the philosophizing of Wittgenstein and Austin which we ignore at our peril. Indeed, it is precisely that complimentarity which permits Hannah Pitkin to draw upon the insights of both in her efforts to understand the fundamental aspects of political life.[34] On the other hand, we should not gloss over the fact that an Austinian concept of philosophy can issue in the notion that philosophy ought to be a *theory* of how we do things with words, a notion which Wittgenstein totally rejects. For Wittgenstein philosophy can only *show* us what everybody knows before they become theorists. It cannot explain anything.[35] Its fundamental task is that of providing a 'perspicuous representation' of our practices, which satisfies our demand for explanation without attempting to theorize about what can only be shown. His aim was to *direct our attention* away from how we *think* about 'thinking', 'intending', 'expecting', 'knowing' and the like and towards what we *do* whan we engage in these activities.

One of the strongest claims in Rubinstein's treatment of Wittgenstein is the claim that there is a quasi-dialectical transition from philosophy to sociology in Wittgenstein's thought, which parallels that in Marx. Getting to the bottom of this claim will involve not simply getting straight about the relation between philosophy and sociology from a Wittgensteinian perspective but also amplifying the point that philosophy cannot be a theory on his view. Rubinstein asserts, "philosophy cannot solve its own problems because ultimately they are not philosophical problems".[36] He goes on to assert, "philosophy must be supplanted or displaced by a kind of social science".[37] He is right to insist that Wittgenstein sees the solutions to those problems which perplexed philosophers from Descartes to Russell in understanding how intellectual activities are rooted in 'forms of life'. However, Wittgenstein by no means proposes to produce a *theory* of 'forms of life'. His aim is simply to show in a coarse way how philosophical problems about, say, meaning are resolved by seeing our practices rightly. Philosophy is the activity of assembling reminders for that purpose. This helps to explain why Wittgenstein left the crucial term 'form of life' so vague. He wanted to illuminate traditional philosophical problems not to produce an account of our social practices. The fact that his elucidations of the origins of philosophical problems can be clarified by reference to our social practices may indeed illuminate those practices but this is not Wittgenstein's aim. He is not trying to describe social practices but to show how they constitute the basis of understanding. His aim is to lead us to a moment of *realization* or insight into ourselves. This explains why the aphorism, the question and the thought experiment rather than straight-forward arguments are his favorite philosophical tactics. What interests Wittgenstein about our social practices is not the contingent features of life in any given society, still less is it a general account of the necessary and sufficient conditions for social practices as such, but the fact that nature has so constituted us that the ways in which we speak about activities such as 'knowing' obscure the fact that knowing is a mode of behavior. His task is simply to remind us of that. Thus, Wittgenstein is less interested in characterizing 'forms of life' than he is with gesturing towards them. Moreover, there is no sense in which Wittgenstein's resolution of philosophical problems in terms of 'forms of life' *requires* either Marxism as its compliment or a transformation of philosophy into sociology. However, to have made this point is to have said something about Wittgenstein; the question of the relevance of the notion of 'forms of life' to sociological explanation remains to be discussed.

We are now at the point in our discussion where we must raise the most important question of all: what would a Wittgensteinian sociology look like? In order to answer this question we shall have to subject the all-important notion of 'forms of life' to close scrutiny. When we have done so we can explore its implications for sociology and, finally, cast some light on the relationship of Wittgenstein to Marx.

Hannah Pitkin has raised the question of what a Wittgensteinian political theory would look like. It will be profitable to take her response to that question as our point of departure. She rightly emphasizes that a Wittgensteinian political theory would, above all, not be a theory but an anti-theory.[38] It would have a strongly casuistic flavor, i.e., insights into the nature of political phenomena would be obtained from meticulously examining individual real and imaginary cases. Further, it would be conflict oriented. Finally, it would be anti-moralistic to the extent that the Wittgensteinian 'theorist' would have the perspective of a participant in, rather than an observer of, political life. Thus, it would proceed from the assumption that everything we need to know about politics is available to political agents, i.e., not a matter of esoteric lore. Thus, the theorist's task would be to assemble reminders of those of our practices which our very involvement on political life obscures.

As far as Pitkin's description goes, it is commendable. However, she does not emphasize nearly enough in developing her picture of Wittgenstein the anti-systematic philosopher that no small part of the thrust of Wittgenstein's philosophizing is directed at showing us how our language contains a tendency to lump together widely diverse phenomena under a single term. All of Wittgenstein's philosophizing, early and later, is directed at driving this point home to us. If it is right, the first task of a Wittgensteinian analysis of politics would be to establish the multiplicity and diversity of political phenomena — 'political games', if you like. In addition, it would emphasize how the normal 'political scientist's' esoteric concept of politics as party politics *conceals* a variety of more homely but, nevertheless, important political phenomena. It would be an exercise in self-awareness to the extent that a Wittgensteinian political philosophy would show us how we are involved in political activities at home, at the job, in cultural debate, in our social relations, etc. without being aware of it. John Danford is right to see in the implications of Wittgenstein's thought for political theory a long-overdue rehabilitation of the Aristotelian notion that human action as such is political.[39] Further, a Wittgensteinian political philosophy would be anti-whiggish to the extent that it would emphasize how different 'political games' are rooted in different 'forms of life'. We need do little more than substitute the word 'social' for 'political' to see the implications of these ideas for Rubinstein's thesis.

However, to have made these suggestions is to brush more closely up against the thorny question of just what a social science oriented around 'forms of life' would be like. The difficulty of this task arises from the obscurity of Wittgenstein's text in the *Philosophical Investigations.* Indeed, the common understanding of the term among social scientists, i.e., as a way of formulating a cultural relativism, has been challenged by Newton Garver.[40] Garver maintains that a close reading of the *Investigations* yields the view that there is one and only one human form of life. The argument is largely philological. In four of the five occurrences of the term in the *Investigations* it is used in the singular. Garver believes that Wittgenstein uses the term to contrast the way human behavior is rooted in man's biology with the ways in which the behavior of apes, dolphins, whales and fish is rooted in their respective biological structures. Garver rightly points out that we only understand 'forms of life' in terms of 'language games'. He also rightly indicates that language games are not to be understood as diverse contexts analogous to dialects in natural languages but are in fact more like geological cross-sections of rock which permit us to observe its striation. Thus, they simplify the complexity of natural language for the sake of displaying the diversity of things which fall together under the rubric of 'language'. If Garver is right, the whole project of basing a social science upon the notion of 'forms of life' involves a serious misunderstanding.

These objections, however, turn out not to be fatal ones, even if Garver should turn out to be correct in his construal of the notion of 'form of life' in Wittgenstein's text. They must, nevertheless, be taken seriously, if for no other reason than to emphasize the difficulties involved in 'applying' Wittgenstein to the social sciences. The proper response to Garver's argument is the following. In the first place, there is no necessity for assuming that Wittgenstein had an articulated notion of form of life, as we have seen. Secondly, in other texts such as the *Lectures and Conversations on Aestethics* Wittgenstein clearly does use the term with just the sort of connotation that social scientists read in it.[41] Thirdly, and most importantly, Wittgenstein's insistance that all seeing is 'seeing as', ultimately rooted in the application of different rules, insures that the requisite relativism is indeed present in the later Wittgenstein.[42] So, regardless of whether the term is correctly employed in these efforts to establish the relevance of the later Wittgenstein to cultural relativism the basic move is justified. Indeed, while Wittgenstein emphasizes that the crucial notion of 'language game' refers to a simplification of ordinary language, there also seems to be some justification in construing language games as local usages. For example, when Wittgenstein suggests that meaning is tied to learning it would seem to follow that the language games according to which we learn the meanings of terms for the first time would be roughly equivalent to local usage.[43] In any case, there is plenty of evidence that Wittgenstein was a committed cultural relativist to the point that the views normally taken to be expressed in shorthand form in the term 'form of life' can legitimately be attributed to Wittgenstein.

However, this tells us precious little about how we are to study 'forms of life' as sociologists. A social science oriented around 'forms of life', it seems, would have to proceed to look at the ways in which societies are constituted by language games, i.e., it would start from the assumption that 'society' is a reification of widely diverse phenomena. It will approach these phenomena by asking how it is that characteristic modes of expression, words and gestures, incorporate different modes of socialization. In short, a Wittgensteinian sociology will involve drawing a kind of semantic map of society, whose aim will be to show us how complex the notion really is. Long ago John Dewey observed, "Society is one word, but infinitely many things. It covers all the ways in which by associating together men share their experiences, and build up common interests and aims; street gangs, schools for burglary, clans, social cliques, trade unions, joint stock corporations villages and international alliances".[44] However, this part of his pragmatist effort at reconstruction in philosophy remains to be carried out. A Wittgensteinian sociology promises to do justice to the complexity of the divisions which constitute society. This certainly compliments Marx to the extent that his great sociological break-through was precisely an account of the way that economic activity results in the fragmentation society.

The particular merit of a Wittgensteinian sociology is that it leaves *open* the question of just *how* society is divided. Indeed, this becomes the central issue in such a sociology. The Wittgensteinian sociologist will proceed from a modest skepticism with regard to the unity of a culture or a class. He will try to discover those linguistic usages which at once bind and distinguish social groups. He will be continually on the look out for the ways in which our characteristic modes of self-description conceal our actual social relations. Such a sociology will conceive society as constituted by agents whose linguistic habits identify them but whose number and nature remain to be identified. Such an approach is predicated upon the notion that shared usage indicates shared modes of rule-following but leaves the question entirely open as to how many modes of socialization are present in a given society. Similarly, it would not pre-judge the question of the compatability of membership in different groups nor the question of the static or dynamic nature of/culture sub-culture distinctions. At the very center of its focus will be the question of disputed usage, i.e., what William Gallie termed 'essential contestability'. The Wittgensteinian sociologist will take overt and covert disputes over meaning as constitutive of social relations. Above all, he will study *struggle*.

We are finally at the point where we can make some suggestions about just what Wittgenstein has to say to Marx. Pitkin is right to assert that what distinguishes Wittgenstein from Marx, Freud, Nietzsche and other thinkers who aimed at forcing us to take a hard, less than pleasant look at ourselves, is the fact that Wittgenstein has neither vision nor program. He simply wants to put us in a position to see ourselves as we really are and not as we would like to be. To be sure this element is by no means lacking in Marx. It explains why a 'revolutionary' would devote the greater part of his time and energy to writing huge, virtually incomprehensible tomes filled with painstaking social and economic analyses. Perhaps Marx's concern that his own followers see the world as it is, and not as they would like it to be, is best expressed in his admonition that men do not make their own history *arbitrarily*.[45] However, Marx is indeed dominated by a vision — and by an assumption. That vision is the vision of a truly free society in which a person has the opportunity to develop whatever talents nature has endowed him with. This requires a radical transformation of society, not merely a change of regime. The assumption upon which this radical transformation of society rests in the union of philosophy and the proletariat.[46] The basic idea behind this union is the notion that philosophy establishes that vision but it requires material interest before it can be transformed into social reality. Marx saw the proletariat as the vehicle for the realization of philosophy precisely because it is totally denuded of any interest except brute survival and thus open to 'formation' by philosophy. Marx assumes that the forces at work in capitalist society tend to universalize proletarianization. Thus the rapacious cut-throat capitalist drives the ever-increasing proletariat into the arms of the philosopher. However, this has not happened; the universal proletarianization which would be the necessary prerequisite for the union of philosophy and the proletariat has not occurred. The world does not divide in the neat way that Marx suggested it would. For all the depth of Marx's analysis he was only able to get a glimpse of the enormous complexity of social relations.

The promise is, then, that Wittgenstein's work, rooted as it is in a cultural pessimism, which is the exact opposite of Marx's optimism, may help us to obtain a more accurate picture of the order within the 'messier' world which we confront. If the view of Marx presented above is correct, then his fundamental inadequacy is in describing the world. If the radical transformation of society which Marx sought is at all possible, it will only be possible to the extent that we have a grasp of the existing order of things. The concreteness of a Wittgensteinian social philosophy promises and advance in the direction of increasing our awareness of the multiplicity and diversity of social phenomena, which will at least help us to grasp why the Marxian vision has *not* materialized. Perhaps in the course of doing so, this social philosophy, whose depth comes from leaving things as they are, *may* illuminate the real possibilities for that transformation of society for which Marx so carelessly worked. This remains to be seen.

Doubtless those consumed by the Marxian vision will find this suggestion wholly unsatisfactory. Nevertheless, for all its cultural pessimism, Wittgensteinian sociology will focus upon those disputes over meaning which so divide society and thereby illuminate the problem of why our efforts to transform society on a grand scale have been less than successful. Above all, the quietist, disinterested perspective Wittgenstein demands of the philosopher — in some respects similar to that which Karl Marx, the social scientist (as opposed to the radical) who actually worked in the British Museum — insures that, whatever implications for political action emerge, there will be no question that there will be any role whatsoever for an intellectual elite in political life. Indeed. it opens up the possibility that the philosopher can play a role analogous to the one Marx himself played in transforming the conspirators of the League of the Just into the educationally-oriented League of Communists, i.e., the role of critic *within* the framework of a movement to transform society. The experience of Marxist movements throughout the world indicates that this is crucially necessary if they are not to be merely political factions but genuinely emancipatory in character. If the implications of Wittgenstein for Marx are sobering, they are also revitalizing. If it seems scandalous to mention Wittgenstein and Marx in the same breath to both Wittgensteinians and Marxists, this mere fact should not be taken to imply that they have nothing to say to each other. Indeed, it may be all the more reason to explore their implications for one another.

NOTES

1 Ludwig Wittgenstein, *Philosophical Investigations*, trans. G. E. M. Anscombe (Oxford, 1969), **I**, 124.

2 Wittgenstein, *Ibid.*, **I**, 128; cf. **I**, 599.

3 J. C. Nyíri, "Wittgenstein's Later Work in Relation to Conservatism", *Wittgenstein and His Times*, ed. B. F. McGuinness (Chicago, 1982), pp. 44—68.

4 Ludwig Wittgenstein, *Tractatus Logico-Philosophicus*, trans. D. F. Pears and B. F. McGuinness (New York, 1961), 4.111.

5 Anthony Kenny, "Wittgenstein on the Nature of Philosophy", *Wittgenstein and His Times*, p. 7.

6 Ludwig Wittgenstein, *Vermischte Bemerkungen*, ed. G. H. von Wright (Frankfurt/Main, 1977), p. 117.

7 Kenny, *op. cit.*, 21

8 Fania Pascal, "Wittgenstein: A Personal Memoir", *Wittgenstein: Sources and Perspectives*, ed. C. G. Luckhardt (Ithaca, N.Y., 1979), p. 25.

9 Wittgenstein, *Investigations*, p. x.

10 Maurice Drury, *The Danger of Words* (London, 1973), p. ix.

[11] Ferrucio Rossi-Landi, "Towards a Marxist Use of Wittgenstein", *Austrian Philosophy: Studies and Texts*, ed. J. C. Nyíri (Munich, 1981), p. 145.
[12] David Rubinstein, *Marx and Wittgenstein: Social Praxis and Social Explanation* (London, 1981).
[13] *Ibid.*, p. 23.
[14] *Ibid.*, p. 24.
[15] *Ibid.*, p. 77.
[16] Alasdair MacIntyre, *Marxism and Christianity* (New York, 1968), p. 88.
[17] Rubinstein, *op. cit.*, 4; cf. Allan Janik and Stephen Toulmin, *Wittgenstein's Vienna* (New York, 1973), p. 227 *et passim*.
[18] G. H. von Wright, "Wittgenstein in Relation to His Times", *Wittgenstein and His Times*, p. 109.
[19] Rubinstein, *op. cit.*, p. 104.
[20] Charles Taylor, "The Opening Arguments of the Phenomenology", *Hegel: A Collection of Critical Essays*, ed. Alasdair MacIntyre (Garden City, 1972), pp. 151—87.
[21] Martin Heidegger, *Sein und Seit* (14th printing; Tübingen, 1977), p. 120.
[22] Richard Rorty, *Philosophy and the Mirror of Nature* (Princeton, 1979).
[23] Hannah Pitkin, *Wittgenstein and Justice*, (Berkley, 1972), p. 326.
[24] *Ibid.*, p. 340.
[25] *Ibid.*, p. 328.
[26] See note 11.
[27] Rossi-Landi, *loc. cit.*
[28] *Ibid.*, p. 139.
[29] J. C. Nyíri, *op. cit.*, p. 58.
[30] Stephen Toulmin, "Ludwig Wittgenstein", *Encounter* **33** (1969), p. 59.
[31] John Passmore, *A Hundred Years of Philosophy* (London, 1957), p. 449n.
[32] Personal Communication from Stephen Toulmin.
[33] Ludwig Wittgenstein, "Remarks on Frazer's *Golden Bough*", trans. John Belvershuis, *Wittgenstein: Sources and Perspectives*, p. 76.
[34] Pitkin, *op. cit.*, p. ix.
[35] Wittgenstein, *Investigations*, **I**, 126.
[36] Rubinstein, *op. cit.*, p. 137.
[37] *Loc. cit.*
[38] Pitkin, *op. cit.*, pp. 325—28.
[39] John Danford, *Wittgenstein and Political Philosophy* (Chicago, 1978), p. 153.
[40] Newton Garver, "Lebensformen", Lecture to the Institute for Philosophy at the University of Graz, November 4, 1982. Unpublished.
[41] Ludwig Wittgenstein, *Vorlesungen und Gespräche über Äthetik, Psychologie und Religion*, trans. Eberhard Bubser (Göttingen, 1968), p. 32; cf. 29.
[42] Wittgenstein, *Investigations*, II, xi, pp. 193—211.
[43] *Ibid.*, **I**, 77.
[44] John Dewey, *Reconstruction in Philosophy* (New York, 1950), p. 156.
[45] Karl Marx, "The Eighteenth Brumaire of Louis Bonaparte", The Marx-Engels Reader, ed. Robert C. Tucker (New York, 1978), p. 595.
[46] Karl Marx, "Towards a Critique of Hegel's Philosophy of Law: Introduction", *Writings of the Young Marx on Philosophy and Society*, ed. Lloyd D. Easton and Kurt H. Guddat (Garden City, 1967), p. 264.

CHAPTER IV

ON EDIFICATION AND CULTURAL CONVERSATION: A CRITIQUE OF RORTY

Being a philosopher today in the wake of the shipwreck of Neo-Kantianism and other such foundationalist ventures is for Richard Rorty a matter of entering into a cultural conversation. The aim of this conversation is to edify in the sense of finding new, better, more interesting, more fruitful ways of expressing ourselves and coping with the world.[1] Rorty alleges that this concept of philosophy emerges from the thought of Wittgenstein, Heidegger and Dewey as they demolished the idea that the goal of philosophy is to represent the world as accurately as possible. For Rorty this does not mark the end of philosophy but the transformation of its role as a discipline. From being a kind of super science philosophy has been transformed into a forum for cultural debate in the hands of these thinkers. However, participation in this conversation requires a special kind of philosophizing, which Rorty designates as edifying. He identifies this mode of philosophizing on the basis of a set of distinctions inspired by the work of Thomas Kuhn. Thus, he distinguishes normal from revolutionary philosophers on the basis of whether a thinker participates in well-established communal efforts to solve philosophical problems or tries to alter the direction or the basic methods of the discipline. Then he goes on to distinguish systematic philosophers from edifying ones. The former are principally interested in making a contribution to a cooperative, progressive discipline; whereas the latter try to dissuade us from doing precisely that. A host of difficulties lurk in these claims and distinctions.

First off, it seems that edifying philosophy is going to have to become normal, if the role of the discipline is being transformed. But how are edifying philosophers to be *educated* to think anti-systematically and to react against what has been taken to be normal? Rorty does not even consider the problem. He blithely alleges that the philosopher's entree into cultural conversation is based upon his skill in analyzing arguments. However, this is a skill which is eminently normal among normal philosophers. It seems that normal and even systematic philosophy is a prerequisite for being edifying; yet receiving such a training would seem to prevent the fledgling philosopher from developing precisely that radical perspective which would allow him to challenge entrenched systematicity. Rorty has not considered this dilemma, I think, principally because he has allowed Thomas Kuhn to do his sociology for him without the slightest idea of the myriad difficulties that this entails. If I am right this cavalier attitude — as for the sociology, our servants can do that for us! — comes back to plague him when he develops his scenario for the future of his transformed discipline.

The second major difficulty with his thesis is that there are grave reasons for doubting that the philosophical programs of his three heros are compatible with one another. Indeed, I shall argue that their differences from one another are such that they cannot be considered to justify a program for edifying philosophy à la Rorty and that their reasoning about what Rorty terms edification is the basis for that incompatibility.

Contrasting Wittgenstein and Heidegger with Dewey will establish that the latter alone fits the paradigm of philosophy as edifying in Rorty's sense of extending cultural conversation. For Dewey alone it is appropriate to speak of post-epistemological philosophy as involvement in cultural debate. This is because Dewey alone takes philosophy to be something *essentially* social and public.[2] In Dewey alone do we find optimism with respect to philosophy's capacity to play a role in promoting social progress. Dewey's optimism in these matters contrasts sharply with Wittgenstein's cultural pessimism, according to which progress is more illusory than real (though not entirely illusory). It is no less discordant with Heidegger's view that we are at a point in history were we lack precisely that shared vocabulary which would facilitate Rorty's cultural conversation.[3] In contrast to Dewey Heidegger sees us in a surrender situation in which all possibilities are opened up to the solitary individual. For Wittgenstein philosophy can only dispel illusions; whereas for Heidegger it gives way to a genuinely edifying *meditation* upon language as the absolute presupposition of anything whatsoever. For all their insistence that meaning is essentially social (more apparent in Wittgenstein than in Heidegger but nevertheless central in the latter's thought), philosophical activity is pre-eminently the activity of an isolated individual for both thinkers. They both justify this on the grounds that to take part in a conversation is to make assumptions, something which brings the philosopher into conflict with his role as laying bare the assumptions that lurk concealed in discourse. Indeed, Wittgenstein criticized one of his closest friends and colleagues, Frank Plumpton Ramsey, for refusing to challenge the assumptions of contemporary mathematics. In a startlingly radical turn of phrase Wittgenstein remarked that this was nothing less than "bourgeois" and, as such, reprehensible.[4] On Wittgenstein's view insight into philosophical problems is only possible from a stance outside of all human communities. This is a view that is congenial to Heidegger because, like Wittgenstein, he assumes that it is precisely our involvement in acting in whatever way we do that prevents us from obtaining a clear perspective upon the actions we are performing. Now, it is less that this perspective on action is foreign to Dewey than that his main interests in action lay elsewhere, i.e., in changing the order in the world, that leads him to de-emphasize the importance of this difference. But, for all that, it is a difference which makes a difference. For Dewey it is above all the way things are in the world that is of paramount importance; whereas for both Wittgenstein and Heidegger the very existence of the world is what philosophy must fathom. Thus, Rorty may be justified in appealing to Wittgenstein and Heidegger as demolishing the traditional epistemological project but he cannot appeal to them as inviting us to take part in a cultural conversation. Nothing belies the differences between the thinkers in question more than the differences in their respective styles of writing, which are considerably more noteworthy philosophically than has hitherto been perceived.

Dewey's works are treatises which argue for theses in much the same way that the foundationalists he so deplores do. In form, however, the differences are minimal. If the difference between Dewey and his adversaries had to be identified, one might call attention to the fact that the *voice* of Dewey's writings is always urgent rather than scholarly; it is eminently political rather than scientific, impassioned rather than detached. However, they by no means differ from traditional philosophical style in the radical way that the writings of Wittgenstein and Heidegger do. The stylistic differences between Wittgenstein and Heidegger on the one hand and Dewey on the other are not merely accidental and external, but bear upon fundamental distinctions in outlook which separate both of them from Dewey and draw Rorty's concepts of cultural conversation and edifying philosophy into question. Both Wittgenstein and Heidegger philosophize *monologically* (for the sake of argument lets say that Heidegger's later work is philosophy, and that in the sense that Wittgenstein's activity is one of the legitimate successors of what used to be called philosophy). Their works contain arguments but they are very peculiar ones. They certainly do not aim at establishing the necessary and sufficient conditions for phenomena through any kind of analysis. In both cases the point of philosophizing is to make us reconsider the assumptions upon which modern philosophy rests with a view to undermining those assumptions. To reconsider basic assumptions, as their all-but-forgotten contemporary, R. G. Collingwood, brilliantly argued in his *Essay on Metaphysics*.[6] is to call the very basis of our questioning into question, i.e., to raise doubts where there was only certainty (Collingwood's "absolute presuppositions" are *ex definitione* what is unquestioned in any intellectual enterprize). Dewey does this too but in the form of an assault from without, as it were. Wittgenstein and Heidegger, on the contrary, do this from *within* the context of Husserlian phenomenology and Russellian analysis respectively. They invite us to rethink the assumptions of, say, intentionality or atomism by forcing us to reflect upon them. They do this by putting us into a position where we must question what we have hitherto taken for granted. In short, they invite us to soliloquize, i.e., to enter into conversation with *ourselves*. Their Socratism results in a new understanding of the sort that each person must attain for himself. This follows in both cases because the questions which are being raised are *transcendental* ones. To assert this is to say that they are questions concerning the relationship of an agent to his own action in a context where agent and action are mutually interdependent in such a way that the investigator turns out to be the subject of investigation. This fact precludes the possibility of proceeding in the manner of science (Dewey's model for philosophy); for I cannot be simultaneously both the investigator and what is investigated. However, I can obtain a perspective according to which I come to have a certain insight into my own situation. This scenario is not always pellucid in Wittgenstein and Heidegger but it is, for all that, there. Ultimately, both thinkers are convinced that philosophy is not knowlege but self-knowledge. As such, it is useless apart from the struggle to obtain it. Philosophical insight cannot be pre-packaged and parceled out as "philosophical truth" because the moment we attempt to formulate it as theses it ceases to be philosophical. It is not that you cannot try to do this, but that you cannot *succeed*; for the result is not "philosophical knowledge" but a parody of what is to be learned from philosophy. Consequently, discursive prose trivializes, rather than illuminates, philosophical problems. In this respect Wittgenstein and Heidegger both differ profoundly from Dewey. Thus, the stylistic differences between them are anything but irrelevant to undeistanding how these thinkers relate to one another.

All three are indeed edifying thinkers in Rorty's sense but Dewey's optimistic mode of edification is crucially different from Wittgenstein's cultural pessimism and Heidegger's suspension of judgment upon such issues. Dewey is interested, like his mentor Hegel, in constructing a truly liberal society. Wittgenstein and Heidegger take the philosopher's withdrawl from society to be the prerequisite for genuine philosophizing. Wittgenstein, for example, insists that philosophy can only understand the world; it cannot change it. Indeed, he goes far enough to insist that the world is understood only to the extent that the philosopher withdraws from it. He once expressed the idea that he would rather see a change in the way people lived, which made the sort of thing he did superfluous, than a continuation of his work — (this, incidentally, is probably why his work has been so difficult to continue). In any case, it is certain that he had no faith in the philosopher's ability to bring about such change. How, after all, can an individual alter the way people become enculturated? To do this would be to control, not only the way parents raise children, but also the *circumstances* under which they raise them. Wittgenstein seems to have thought that cultural conversation in its present from goes as far as it can to obscure this point. Thus his desire to *stop* it.

Heidegger's profoundest belief about this point in history is that we stand in a cultural void in which each individual is left to his own devices to cope as best he can. This is the message he reads in the late Nietzsche. He takes it to be the task of the post-philosophical "thinker", his task, to draw the implications of this fact. We are confronted with the mere fragments of a tradition. We cannot hope to build upon them. As individuals we are left with the choice between vainly resisting this situation in a context where such resistance is an arbitrary effort to fill the gap left by the end of philosophy or we can open ourselves to the *Er-eignis* (the event of appropriation) by surrendering to this condition and, thus, appropriating its possibilities. The correlative to our lack of certainty in the modern world is nothing less than the fact that everything is possible. But everything is possible because everything is questionable. Consequently, anything from self-destruction to radically new modes of creative activity is possible. His later works are little less than a series of mediations upon this condition. Thus, for Heidegger there is no conversation to continue nor are we in a position to initiate one.

To see the full significance of these views and the ways in which they undermine Rorty's thesis we must look more specifically into the ways in which Wittgenstein and Heidegger propose to edify us. We must look particularly at the *tactics* that they employ; for in the differences between their tactics we shall also disscover two *dissimilar* modes of edification.

What are Wittgenstein's philosophical tools? In the *Tractatus*, which, he warns us, we separate from the *Philosophical Investigations* at our peril, he employs aphorisms to get at what propositions cannot say but only show. He uses aphorisms to tease from language that which cannot be expressed in the form of a definite description. His message is paradoxical, namely, that there are subjects such as God, the sense of life, the Mystical etc., about which we cannot speak. It is important to point out that he is well aware that we want to speak about these matters and that we construct sentences about them. His point is that we can never produce anything but nonsense with them. By keeping silence in these matters we leave things as they ought to be. His technique for driving home this realization is indirect: it involves inducing us to reflect upon just what it is to say something with *precision* (i.e., something that is not nonsense) in the first place. We learn that we can say things because language has a logical structure which both facilitates its application and can be made to *show* the sorts of things which cannot be subjects of assertion. He accomplishes this aphoristically by forcing us to take a second look at just what it is we *do* when we use logic.[7]

In the *Philosophical Investigations* he also employs aphorisms along with questions, jokes, examples thought experiments and analogies to bring his reader to an awareness of the ways in which we are constrained by language, which is not so radically different from that of the *Tractatus* as Rorty and others would have us believe. Here his aim is to show us that the problems which have tormented philosophers concerning the nature of language do not require a theory for their resolution but, rather, a proper understanding of the social practices which constitute the order in everyday life. In aid of this Wittgenstein poses to himself — and invites his reader to do likewise — nearly eight hundred questions, only one hundred of which he answers, and then in seventy cases wrongly.[8] His aim, as in the *Tractatus*, is to make us take a second look at what we take for granted about our world. Just as his aim in the *Tractatus* was to get us to stop asking unanswerable questions about God, ethics and aesthetics, so in the *Investigations* he wants to disabuse us of the pernicious habit of asking "philosophical" questions about meaning, knowing, intending and the like with a view to a theoretical answer. Here his goal is to show us how an accurate description of the way things are will allow him — and everyone else — to stop doing philosophy once and for all. He admitted that this was never an easy thing to do because philosophy was not merely a body of pseudo-propositions which answered a body of pseudo-questions, but an ineliminable tendency in human nature as well.[9] His aim was nothing less than providing us with therapies for dealing with the manifestations of this "sick" tendency by giving us a clear view of how philosophical questioning in the traditional sense confuses us about the way things are and ultimately about ourselves.

In the end his aim is to make us see ourselves as we really are and not as we would like to see ourselves. There is no doubt that he anticipated that we would be disappointed by what we discovered. Wittgenstein is certain that we can do nothing about our situation. His work is edifying, then, in the sense that we come to build an accurate picture of the way our practices determine what we become, but it is a far cry from an invitation to join in a cultural conversation which would discover new ways of coping with the world etc. His aim was little less than the abolition of a discipline. This is not to say that the practical skills he utilized as a philosopher would cease to be useful and important. Far from it, he warned one of his favorite students, Maurice Drury, away from a career in philosophy but at the same time told him by no means to stop thinking. However, this is nothing like a cultural conversation. For Wittgenstein to enter into such a conversation is to abandon that very "outsider" perspective, that independence from the standards of the community, which alone makes his peculiar mode of philosophizing possible. Nothing could be farther from Wittgenstein's conception of philosophy than Dewey's liberal, reformist version of pragmatism. This in itself, of course, says nothing about the superiority of his position but it ought to cast doubt upon putting Dewey and Wittgenstein together under a single rubric in the way Rorty has. The same is true for Heidegger.

Heidegger's later thought emerges self-consciously from a tradition of Christian edification.[10] A wholly secularized version of Kierkgaard's brilliant development of the spiritual exercise is at the very center of Heidegger's later work. In Heidegger what I have termed spiritual practices are in aid of conveying a sense of reality for which post-epistemological thought has no vocabulary. Describing this effort at edification and developing the *contrast* between Wittgenstein and Heidegger which emerges will illuminate the dubious nature of Rorty's usage in selecting his metaphor. Heidegger's deepest convistion entails the thesis, not only that Nietzsche announced the end of "Platonism", but that he even transcended it in the "perspectivism" hinted at in *Der Wille zur Macht*. The center of this teaching is the idea that we no longer possess a "paradigm" of rationality as such but only particular paradigms which permit us to pursue particular scientific and social projects in the absence of an over-arching set of principles for co-ordinating life as such and which can, consequently, function as a tribunal for adjudicating disputes between the projects developed from with the various "perspectives".

I have introduced Heidegger in terms of Thomas Kuhn's paradigm language deliberately. I have done this, not because I think Kuhn is especially illuminating — in fact I think that paradigm language obscures the cultural questions people have wanted to illuminate with it — but because the kind of radical transformation in intellectual life which Heidegger alleges has been part and parcel of the end of Platonism involves generalization and universalization of just the sort of incommensurability and incompatibility that characterize Kuhnian paradigm shifts. In short, I think that we might describe the Heideggerian picture of thc current state of culture as "paradigmlessness". What are our options in this situation? In his large study of Nietzsche Heidegger suggests that there are basically two possibilities: arbitrarily asserting wholeness (in the form so developing a *Weltanschauung*), where there is in reality only discontinuity and conflict, or surrendering oneself to the *possibilities* inherent in this otherwise highly threatening situation. His technique for attaining balance in a vertiginous world is in fact a secularized version of the traditional mode of attaining the proper disposition for accepting the demands of religious life, the spiritual exercise. Actually, Heidegger's later writings incorporate the Kierkegaardian development of traditional religious character-forming techniques in his *Edifying Discourses*, which raised the form to unequalled power and effectiveness through the Dane's consummate literary artistry. Since I have discussed this topic at length elsewhere in this volume, I shall not do so here; rather I shall now turn to the consideration of just what possibilities actually exist for cultural conversation in a situation which is similar to but by no means identical with the one Heidegger describes.

The problem with Heidegger (and also with Kuhn) is that he still indirectly is in thrall to logic to the extent that he follows Hegel in viewing philosophy from Thales to Hegel himself as a kind of logically homogeneous unit. In this respect he seems to agree with his adversary Rudolf Carnap in assuming that our various perspectives on reality (i.e., what Carnap unreflectively and superfially considers to be a matter of choice *between* languages) depend upon our behaving logically or consistently once we have arrived inside (chosen) a particular paradigm. The work of scholars such as Richard Popkin and Frances Yates have called the Hegelian picture which has dominated efforts to write the history of philosophy from the global perspective i.e., in the work of Copleston, Bréhier, Windelband etc.) into question.[11] The notion that we are engulfed or encapsulated within our various perspectives on reality has been contested by a philosopher who is conspicuously absent in Rorty's book, Karl Popper.[12] This is puzzling indeed because, of the seminal figures in twentieth century thought, none has insisted more adamantly than Popper that it is possible to communicate without sharing an articulated framework for discussion. However, if Popper has maintained *that* such dialogue is possible in the midst of basic disagreement, he has not explained *how* it is possible to carry on conversation in the absence of commonly recognized principles (i.e., in a situation in which each party to the conversation is liable to take his antagonist's view as based upon assumptions which are wholly unwarranted). Wittgenstein, it will be remembered, insisted upon the fact that communication rested upon agreements in *judgment*. If there is anything at all to the claims that Heidegger has made about "the era of paradigmlessness" (die Zeit des Weltbildes) — and the dazzlingly brilliant efforts of Alasdair MacIntyre to sort out the sources of contemporary alienation in the logic of our "absolute presuppositions" about morality indicate very strongly that there is[13] — Rorty has no right whatsoever to invite us placidly to participate in a conversation that is more than likely to be *pointless* (yet another mark of the superficiality of Rorty's concept of cultural conversation is reflected in the fact that he has not even considered the implication of John Austin's notion that it is the adverbial modifiers of actions that are philosophically interesting for it). In short, there is no lack of evidence to indicate that those very cultural dilemmas which press the sort of cultural conversation Rorty envisions upon us almost insures that this conversation will quickly degenerate into a chaotic shouting match. This possibility seems not to have crossed Rorty's mind. In his enthusiasm for establishing a role for edifying philosophy in our cultural debates Rorty has overlooked the fact that our actual public debates — think of abortion, equality and nuclear power — are at loggerheads. If MacIntyre is right, this is not accidental. It seems then, that in his pursuit of his Deweyite ideals he has all but ignored the communicative context with which our society presents him. However, to assert that Rorty failed to recognize the depth and significance of social and intellectual divisions and how these divisions militate against his ideal cultural conversation is not to assert that such dialogue is altogether impossible. In fact, this situation has been a matter of concern for a number of philosophers. I should like to round out this discussion of Rorty's thesis by providing an outline of a sketch of what I take to be the possibilities for cultural conversation in our world. I shall concentrate a notable effort to establish just how conversation can go on in the absence of a shared framework for discussion.

In one of the most important papers in the philosophy of culture in this century. "Essentially-Contested Concepts",[14] William Gallie has brilliantly shown that the deep-seated disagreements concerning fundamental issues in art, politics or religion need not be construed as irrational conflicts between different sets of emotional commitments but can be rational in nature. Gallie's achievement is to have provided us with criteria for discerning whether or not such deep-seated disagreements do indeed have a rational foundation. His basic idea is that conversations can arrive at an impasse legitimately. This happens when parties to the conversation rest their reasoning upon different criteria for determining what falls under a concept. This can happen legitimately because concepts like art, Christianity, democracy etc. are highly complex. Their complexity permits them to be construed in more than one way, i.e., each mode of construal emphasizes the centrality of different features as "defining characteristics" of the concept in question. Gallie's aim was to establish criteria for distinguishing those discussions which come to loggerheads legitimately over matters of essential contestability from those which arrive at that point as a result of radical confusion. The details of his analysis are too complex to be discussed here. Suffice it to say that this argument rests upon the notion that there are different traditions of interpretation of concepts of a complicated, important and historically long-lived phenomena, which are such that they can recognize common authorities (as all Christians recognize the authority of, say, St. Paul). If there are different traditions which are well-established, there are legitimate grounds for disagreement and we can at least come to grasp how and why we disagree. This permits us to agree to disagree and leaves some possibility for continuing our conversation in the face of deep-seated, basic disagreement. To be sure, this procedure for determining each other's canons of reasoning and the traditions in which they are rooted by no means assures that future agreement will emerge but at least that possibility remains open. On this view, then, we agree to disagree on the basis of articulating our mutual understanding of tradition as well as the specific precedents upon which our reasoning rests. The conversation is kept going because we are both interested in getting to the bottom of the issue. Thus, the hope is that by coming to understand each other's reasoning we *may* be able to uncover hitherto concealed possibilities for discourse.

William Connolly has drawn a very important corollary from Gallie's work,[15] which has widespread implications for Rorty's notion of cultural conversation. Connolly insists that to discover that a concept is essentially-contested is to discover that we are in the midst of a *political* contest. Indeed, Connolly goes so far as to insist that essential-contestedness is the defining characteristic of political life. When we use an essentially-contested concept we are exercising power. If this is right, then Rorty's cultural conversation will be political in nature. That is to say that we should expect tempers to flare at worst and at best to have to expend large amounts of time in patient persuasion. Although Rorty has consciously borrowed the term cultural conversation from one of this century's leading political philosophers, Michael Oakeshott, he seems oblivious to its political dimension. Unlike Dewey, Rorty seems unaware that he is asking that philosophy as a discipline becomes overtly political — in Connolly's Aristotelian sense rather than in the sense of partisan politics. This is certainly not a bad thing; for philosophy always was political in the sense mentioned. We can only benefit from bringing that dimension into the open at long last. If philosophers want to follow Rorty's suggestions about becoming edifying thinkers, they shall have to be prepared to enter an overtly political forum. Since Marxists, feminists and other radicals — as well as not a few conservatives — suggest that there are good reasons for thinking that the authors quo is imbued with covert politics, making philosophy overtly political in the Aristotelian sense mentioned can only be a step forward in a genuinely Deweyite reconstruction of the subject. Thus, it is less that there is anything inherently objectionable about the politicization of philosophy that Rorty unwittingly advocates, but that he certainly should have been aware of the consequences of his suggestion to turn philosophy into an edifying discipline that is of prime importance here.

The point of all of this has been a critique of the naiveté implicit in Rorty's notions of edifying philosophy and cultural conversation. It has not been to disparage his goals but to show that the thinkers upon whom he bases his position and the very logic of the concepts involved requires a different, fuller analysis than Rorty has provided. If I am right, this gap will only be filled when we recognize the logical problems that confront contemporary cultural hermeneutics — and the eminently *political* character of the conversation to which we have been invited.

NOTES

[1] Richard Rorty, *Philosophy and the Mirror of Nature* (Princeton. 1979), p. 359.
[2] See John Dewey, *Reconstruction in Philosophy* (New York, 1950), *passim.*
[3] Cf. "Style and Idea in the Later Heidegger", above.
[4] Anthony Kenny, "Wittgenstein and the Nature of Philosophy", *Wittgenstein and His Times*, ed. B. F. McGuinness (Chicago, 1982), p. 21.
[5] Ludwig Wittgenstein, "Lecture on Ethics", *Philosophical Review* **74** (1965), 8—9; Martin Heidegger, *Einführung in der Metaphysik* (Tübingen, 1958), *passim.*
[6] R. G. Collingwood, *An Essay on Metaphysics* (Oxford, 1940), p. 31.
[7] On the aphorism as a philosophical tactic see J. P. Stern, *Lichtenberg* (Bloomington, 1959).
[8] K. T. Fann, *Wittgenstein's Conception of Philosophy* (Berkeley and Los Angeles, 1969), p. 109.
[9] Kenny, *op. cit.*, p. 25.
[10] See note 3.
[11] Richard Popkin, *The History of Skepticism* (New York, 1968); Frances Yates, *The Art of Memory* (Chicago, 1966).
[12] Karl R. Popper, "Wie Ich die Philosophie Sehe", *Conceptus* **XI** (1977), p. 17.
[13] Alasdair MacIntyre, "How Virtue Becomes Vice", *Encounter* (July, 1975), pp. 11—7.

[14] W. B. Gallie, "Essentially-Contested Concepts", *The Importance of Language*, ed. Max Black (Engelwood Cliffs, 1962), pp. 121–46.

[15] William Connolly, *The Terms of Political Discourse* Lexington, Mass, 1974).

CHAPTER V

TOWARDS A WITTGENSTEINIAN METAPHYSICS OF THE POLITICAL

What is political? The question is stupefyingly simple. It leaves you dumbfounded. Like St. Augustine, everybody knows till they are asked but when asked are unable to reply. Political "scientists", above all, avoid it like the plague. They are quick to point out how silly, how "philosophical" (i.e., naïve/unimportant), it is to raise such an issue. Precious few of them are prepared to tackle it. Radicals such as certain feminists and Marxians (not Marxists), as well as some conservatives, on the other hand, suggest to us that our chief political problems arise from an unquestioning acceptance of the status quo; which is precisely because we fail to pose just such silly questions (R. G. Collingwood's view of metaphysics is one which takes posing such silly questions to be metaphysics *par excellence*). To fail to pose this question in the minds of these critics of the existing order is to put yourself in a lamentable situation where you remain blind to — and, therefore, coerced by — class structure, patriarchy or trendy notions about, say, egalitarianism. When we assume, they continue, that politics is merely a matter of "public life", i.e., of office-seeking, law-making, dictating public order, and, therefore, that the subject that the study of politics concerns itself with is the means of obtaining and maintaining the power to do so, we obscure the economic forces, family structures or "social ideals" which dictate the nature of existing political arrangements. In short, by accepting the common understanding of what it is to secure and retain power we systematically blind ourselves to alternatives. However, radicals and conservatives both have a penchant for putting their message so sanctimoniously that the important philosophical question they intend to pose is itself obscured by their zeal. Their problem is in fact the obverse of that of conventional politicians and political analysts of the right and the left, who refuse to admit that our question is remarkably difficult and no less important. Both radical and conservative proponents of alternatives to the status quo and its liberal defenders, right and left, want to use political knowledge as a *weapon*. The point of posing the question "What is political?" simply and straight-forwardly, *sine ira et studio*, is to obtain a perspective upon the role of The

Political in human experience which will enable us to grasp how politics fits into human practices without at the same time endeavoring either to justify or to condemn existing political arrangements. In their haste to change the existing order radicals have a tendency to assume unquestioningly that it can be changed. In this they are all *too* liberal. Similarly, conservative critics of the status quo are too liberal to the extent that they want to *change* society. Oddly, the defenders of the status quo right and left *fail to be liberal enough* in ignoring the suggestion that the role of politics might be considerably larger than public officials, journalists and social "scientists" assume. It is precisely the aim of political *philosophy* as conceived here to raise the question "what is political?" anew in a way which challenges the assumptions of social "scientists", politicians and mere mortals alike from the perspective of a philosopher who rejected theorizing for the sake of obtaining *clarity* about the way things are by leaving them as they are.

On the face of it there seems to be something very odd about approaching a subject that played no role whatever in Wittgenstein's life or thought from a perspective drawn from his later philosophy. This deserves comment. In the *Tractatus* he sharply demarcated what could be put into words *precisely* (i.e., with absolutely no ambiguity), from what could not. He confused a whole generation of philosophers and scholars by opposing the two as the *meaningful* and the *meaningless*. The former was the subject of science; the latter he designated ethics. Later in his "Lecture on Ethics" he cast a little light upon the obscurities of this effort to consign ethics to silence by distinguishing matters of absolute value from matters of relative value. The former referred to questions about God, the sense of the world and the meaning of life; whereas the latter seemed to refer to things like distributive justice. It is necessary to emphasize that Wittgenstein was so little interested in questions of relative moral value that he never specified what he meant by the term except to say that they were all the questions in ethics which were not matters of absolute value. It is obvious from his text and everything else he wrote as well as the testimony of his lifelong friends that questions of relative value, questions about politics included, unlike questions of aesthetics, did not interest him *in the least*. So, it is indeed strange to discuss the implications of his later philosophy for our understanding of politics. However, we should not forget that Wittgenstein's own mode of describing what he was doing is by no means infallible. For example, what he termed ethics in the *Tractatus*, i.e., questions about God, the meaning of life and the limits of language, have been treated by other philosophers as metaphysics or the meeting point of ethics and metaphysics (as they are, say, in Karl Jaspers). Moreover, the conception of language as crystalline, which facilitated gesturing at the ethical in the *Tractatus* is precisely what is no longer available in his later philosophy, despite the fact that Wittgenstein never revised his views on the matter. Yet, there is much to suggest that he should have. Further, it is just the family-resemblance character of concepts, as William Connolly has suggested and upon which it will be necessary to expand below, that both replaces the neat world of definite descriptions and introduces a political dimension to Wittgenstein's account of concept formation. However, this is to run ahead.

To say that it is strange to discuss Wittgenstein in relation to politics is by no means to assert that it is pointless as an increasing number of scholars have emphasized. Consider these examples. Hannah Pitkin insists that there are important implications for our understanding of political theory in the late Wittgenstein's concept of language as constituted through cues rather than labels as well as in his view of philosophy as leading to self-awareness. J. C. Nyíri has argued that Wittgenstein was in fact a conservative thinker, who devoted himself to developing the philosophical underpinings of conservatism in asserting, for example, that we follow rules blindly. Ferrucio Rossi-Landi, on the other hand, has linked him to Marx in describing him as the analyst of linguistic alienation. For David Rubinstein Wittgenstein can be construed as complimenting Marx's account of practice and, therefore, providing a sounder foundation for social theory than any of the other currently available alternatives. As provocatively stimulating as these latter efforts to associate Wittgenstein and politics have been, they are problematic to the extent that Wittgenstein explicitly wanted to stress how little his work belonged to the age of Fascism and Socialism *both*. He was less interested in erecting structures than he was in examining the foundations of existing ones. To accomplish this he thought that the philosopher had to renounce citizenship in any human community (contra Neurath). If these sentiments are not mere wilfulness — and what follows endeavors to articulate how they are not — they are not only central to his concept of philosophy as *necessarily* having to leave things as they are but also have wideranging implications. The perspective he would obtain upon the world was one which eschewed looking forward or back to see where we are but planted itself firmly in the present (Nietzsche — die ewige Wiederkehr des Gleichens). In short, Wittgenstein endeavored to be as a-political as he could possibly be both as a man and as a philosopher. Paradoxically, it is precisely this a-political stance which constitutes his importance for political philosophy. In his efforts to avoid being blinkered by the ubiquitous ideologies in his world, cultural (e.g. Bloomsbury and the Brenner Circle) as well as political, he developed an eminently anti-ideological mode of philosophizing, whose main resources remain largely untapped. For the sake of those skeptics who find his dictum that philosophy leaves everything as it is a hopelessly unacceptable endorsement of the status quo it should be emphasized that this in no way involved endorsing the legitimacy of the status quo but only that philosophy fails in her task of understanding the world when she directs herself to changing it. How the world gets changed is another matter, one for which Wittgenstein believed nobody had an answer. If this is right, then Wittgenstein dismissed "political" radicalism precisely in order to adopt a yet more radical perspective, one which would allow him to go to the root of the problems which have tormented philosophers for so long. The main lines of his critique of epistemology are well known (even if the centrality of the notion of rule-following has, with the exception of the work of Kjell S. Johannessen, been insufficiently recognized). The question, then, is how does the radical conventionalism of that critique illuminate the question "what is political?"; for on the surface they are worlds apart.

The most interesting effort to do political theory, however tentatively, in Wittgenstein's name to date is that of William Connolly. His brilliant, all-to-little-appreciated, *The Terms of Political Discourse*, deserves special mention for its authentically Wittgensteinian critique of political "science". To act, Connolly argues, is to employ concepts. In the course of acting we sometimes discover that the concepts we employ (his examples include responsibility, power, interest and freedom) are not wholly shared by other members of the community. His merit is to have demonstrated that these conceptual conflicts are the very stuff of political life. He has trenchantly lain bare the role of essential-contestability (first articulated in Gallie's important paper, "Essentially-Contested Concepts") in political struggle and thereby exposed the politics of "politics"; for the very concept of politics turns out to be itself political. Because many of the concepts we employ are complex, rooted in traditions with a branching development and open to yet further modification, they are capable of being construed *legitimately* in different ways. His analysis of the ways in which "freedom", "power", and "interest" are essentially contested, while by no means drawn exclusively from Wittgenstein, are *explicitly* in aid of the Wittgensteinian goal of establishing the "grammar of politics". Thus, while Connolly does not explicitly emphasize the connection between essential-contestedness and the family resemblance character of concepts in an effort which is otherwise explicitly Wittgensteinian, it is at this point that his analysis "attaches" to Wittgenstein's later philosophy. No small part of Connolly's achievement is to have demonstrated that the very act of describing a situation entails pre-judging its political significance in a way that is *wholly* unavoidable. This implies that "political science" cannot be done from a perspective *au dessus de la melée*. He thus makes good Pitkin's claim that part of Wittgenstein's significance for political *analysis* (NB not theory or philosophy) must be unprivileged, written from within political contests and addressed to fellow citizens.

What follows is a sketch of the implications of Wittgenstein's later philosophy for our understanding of politics. Although it owes much to Connolly and others by way of inspiration, it is not political analysis but political philosophy. Connolly's work proceeds from the problem of establishing the conceptual dimension involved in adequately describing political life and draws implications for both political conduct and analysis from it. His merit (although he does not put it this way) is to have demonstrated, not only that there is indeed an identity between Aristotle's two definitions of man, the *zōon politikon* and the *zōon logon echon*, but also that there is a deep connection between Wittgenstein and Aristotle (and a-fortiori Aristotelians like Hegel and Marx). What follows is an exploration of some prominent features of that terrain which proceeds from the question "how, on the basis of Wittgenstein's later philosophy is politics at all possible?". The question may ring Kantian but the effort is Aristotelian; it is the matter of further articulating how a speaking being must *eo ipso* be a political being. It is necessary to reiterate, then, that what follows is political philosophy and not political theory — still less political "science". It is Wittgensteinian in the sense that is seeks "bedrock", i.e., to "show" how "politics" too is rooted in certain general facts of human nature and, therefore, how it forms a part of our natural history. It is exploratory and makes no claim at being definitive.

To know anything at all about Wittgenstein's later philosophy is to know that Wittgenstein insists that a private language, i.e. one in which I create names for my sensory experiences, is a conceptual impossibility. For me to do this would entail not simply correlating names with sense data but a knowledge of the practice of naming. To think that there can be a private language is to forget, as philosophers often have, that language is not simply a matter of labeling experiences but of doing this deliberately. But, in order to do this deliberately, labeling cannot be of the essence of language; for to label deliberately we have to be *trained* to do so. To consider seriously the possibility of a wholly private language is to forget that labeling is an activity that has to be *learned.* This is what is conceptually impossible: it is inconceivable that someone could understand a name without knowing what naming is all about. "Private languages" such as codes are thus parasitic upon public, meaning-constituting practices.

Speaking a language involves mastering a technique or following rules that we did not create. Neither the substance of the rules nor the details of how they are followed matters in establishing this, but simply the notion that the idea of a private language is entertainable precisely because we have smuggled the notion of language in illicitly through the side door as it were. To learn, then, is to learn something, which is to follow rules (master a technique) that you did not create. To use language, then, is in a certain sense, to be *ruled.*

Viewed from this perspective the implications of Wittgenstein's late philosophizing for our understanding of politics follow fast and furious from his most central teaching. But what does it mean to be "ruled" in this sense? For a start, it means to be subject to behavioral regularities which you have neither created nor approve of and which might be different from what they are in substance. It does not take profound reflection to discover that the rules that any given individual has learned do not *have* to be what they are. Rousseau seems to have got it backwards; for it seems that man is born helpless only to become chained in the sense of becoming constrained by rules not of his making which could be otherwise but which, nevertheless, make action possible. Further, this has to be the case; for we cannot learn to learn without learning something specific. Moreover, it is precisely this mastery of technique which makes us to be what we are. Rules, then, are a-priori in that they make phenomena (action) possible. It is precisely the synthetic and a-priori character of rules that make them in Aristotle's felicitous phrase *potentially* political.

Neither Wittgenstein nor anyone else ever suggested that the rules mastered in childhood were anything but contingent as to their substance. By all indications the rules we internalize as we are enculturated could be different. Each of us, for example, could have had a different mother tongue from the one we in fact have. The rub is, however, that to learn anything we have to learn to follow rules *blindly* (as all army sergeants, novice masters and athletic coaches well know). To become your own person, it seems, you have to first become what your guardians want you to be; not in the sense that your personality is established or fixed by the rules you master and how you master them but in the sense that their rules establish the *range* of possibilities open to you. To realize any human potential whatsoever we have to learn language and, therefore, to master techniques which, from the point of view of their stipualtions, are arbitrary and coercive, as we soon realize that they could be different. What we only realize later, if at all, is that our capacity for action is bound to the discipline which facilitates that action at the same time that it constrains. Belonging to another social class, living in a different place, having different parents etc. would make a great difference in what we learned. To know that I am constrained in the sense that I find myself in a situation that could be different but is not. To be "ruled" by rules is to be always potentially in this situation.

On this view, then, Wittgenstein seems to give aid and comfort to those Marxists, Feminists and Conservatives who take the conditions of production, the family as we know it and "ideals" to be inherently political. However, to take this view is to see only one side of the implications of his concept of rule-following; for he emphasizes that the very repetition which makes concept formation possible makes it no less difficult to throw off a concept at will than it is to acquire one that way. On this view to realize that there is an alternative to the existing order says nothing whatsoever about its "availability". The latter is entirely a matter of our practices to accomodate change, which is a wholly different matter. This follows from the fact that, while the substance of the rules we have mastered determines what constrains us politically (i.e. what makes politics in the ordinary sense possible), it is the activity of rule-following which determines our capacity to change our behavior patterns. As the two become detatched from one another the scenario gets more complex as new possibilities for "entanglement in the rules" arises.

Connolly has rightly identified the political character of essential-contestability; what he has failed to do (and this should by no means be taken to count against his work, which has different goals) is to see that politics inheres in the family resemblance character of concepts. He has written, "When groups range themselves around essentially-contested concepts, politics is the mode in which the clash is normally expressed". What is obscured in this sentence is the fact that for a concept to be essentially-contested groups must have *already* in some sense ranged themselves around it; for they have no existence outside of human actions. Wittgenstein's merit is to have called attention to *how* this can happen. Concepts, he insists, do not permit of neat delimitation or definition by genus and specific difference in Aristotle's terms. Rather, they are looser groupings with significant overlappings but lacking in a set of shared "universal" characterisitcs. They are like members of a family who share sub-sets of characteristics without sharing any "essential" set of common characteristics. Wittgenstein's well-known example is that of the concept game. He is anxious to defend the view that games do not share anything in common to be called games — nor should they be expected to. Implicitly he is claiming that on Aristotelian criteria we are never able to define what we commonly group under the heading game. (This is, in itself, something which Aristotle would agree with; his concept of definition is only intended to apply to natural kinds and not to artifacts. The differences between Aristotle and Wittgenstein here run deep indeed.) In fact common usage decides what is a game such that, when the philosopher produces a definition of game, which, say, precludes games played alone from being termed games properly so-called; the objection, "you have omitted patience and a variety of things that mere mortal language users normally term games," must be taken seriously. Wittgenstein insists that it is well-established social practice rather than philosophers who decide what is real.

We call things "games" or "representatives" or "works of art" because they fit into practices such that the analogies between them suggest that they are members of the same family. Thus there are closely related cousins such as baseball and cricket and more distantly related ones like chess and charades or football and guessing riddles. In many instances, as in the case of the invention of basketball, we know exactly what analogies with football were drawn upon to constitute the game. Admission to the family rests upon the role that the activity plays in our practices as opposed to any of the specific characteristics of the game itself (the actions involved in playing chess might well belong to a religious ritual rather than a game and there is nothing about the rules that tells us that chess is a game). However, because a given individual must learn specific usages to learn language and because we can *invent* games on the other (something that never would have occurred to Aristotle as relevant to our understanding of concepts as it might to, say, Protagoras or Isocrates), it is possible to find ourselves in a position of arguing about whether or not taking cabbages from other people's gardens on Halloween, say, is a game. In this case the line between games, pranks, and vandalism is mighty thin. However, to have made this suggestion is to initiate a political discussion about who belongs to the family; for, although our concepts are flexible, that by no means implies that they are arbitrarily used. Just how flexible is our usage as regards the concept game? What kind of games can *we* invent? What can count as "game behavior"? How much do the surrounding practices with regard to property etc. permit us to consider something that would normally be termed stealing a game? Here we are confronted with a classical political issue. Not only is the usage of the term game at stake here but redefinition of a whole series of practices. It is also noteworthy that the considerations which would lead us to accept the putative innovation, i.e., the game-like character of helping ourselves to other people's cabbages in a not altogether malicous way, are largely incommensurable with the kinds of objections which might be raised to such a suggestion. Therefore, the elements which come into play in resolving the question will also be subject to dispute. Thus concepts can become political issues due to their family resemblance character.

The family resemblance character of concepts, in turn, is rooted in Wittgenstein's central insight, i.e., that to have a concept is to have learned to follow a rule. This is to be disciplined in a particular way. It is of the essence of a rule, however, that it be supple, i.e., that there be more than one way of following the rule. This is not to assert that rule-following is arbitrary but that the context in which rules are applied can determine variations in how the rule is interpreted. In all of Europe, for example, one signals with one's fingers "how many", say, when one orders champagne at the opera. However, in England or Norway, as opposed to France and Austria, one counts from the index finger (as opposed to beginning with the thumb as one). To learn to count to one's fingers, then, will be different depending upon where one has learned the practice. Now it is the conventional (arbitrary) element in the practice (as opposed to its intrinsic order) which is potentially political. This potential is actualized when such customs for whatever reasons come into conflict. For this reason "political" debates can range themselves around skirmishes over the borders of concepts, i.e., over just what falls under a concept. Connolly has provided an admirable account of how political contests in the normal sense are inter alia constituted by conceptual disputes; the aim here has been to bring out just how this is part and parcel of language use and a-fortiori human nature.

This is perhaps a good point for a reminder: the aim here is not to provide a comprehensive account of the political role of concepts but to illustrate the ways in which "politics" in the conventional sense depends upon the political nature of concepts (rule-following). The political analyst (e.g. William Connolly) wants to explain how it is that concepts constitute political debate. The political *philosopher*, on the other hand, wants to show why this *has to be the case* given the kind of creature that we are. This it what is means to speak of the philosopher's task as one of determining how politics fits into our natural history. In effect, to draw this distinction is to distinguish between the dimension of grammar (in Wittgenstein's sense of logical gammar) and the place of politics in our natural history — the two are distinguishable even if they are not separable from one another. The matter might be summed up this way: political analysts want to understand how political concepts work; whereas political philosophy aims at answering the metaphysical question. "how is politics at all possible?".

To sum up. The essentially-contested character of concepts is a constitutive element in political contests but more than that it is the very defining characteristic of political conflict. Essential-contestablity is rooted in the family resemblance character of concepts, which explains how it is that there can be essentially-contested concepts in the first place. Concepts have this family resemblance character because (1) to possess a concept is to be able to apply a rule and (2) rules can be applied in a variety of ways. This is because they have to be learned in the context of *specific* practices, which themselves take their character from neighboring concepts (we do not learn isolated judgments but nests of judgments). But all of this in turn implies that the life of a speaking being is shot-through with politics in the sense that the very first skills we acquire are imbued with a contingent and, therefore, potentially political character. Politics, then, does not refer merely to those aspects of public life surrounding law-making but is an ineliminable element in concept-acquisition and use. If this is correct, radical revision of the nature of social science is required in the sense that to understand "politics" we shall have to grasp how it grows out of the political dimension of our practices (languages games). Moreover, it follows that we shall have to make important adjustments as regards our understanding of the ways that politics enters into our lives; for now it becomes tied to our very first learning experiences. Thus interpersonal relations, scientific debates and artistic activities, for example will have to be understood as having a political dimension. Politics, as Connolly rightly emphasizes, pervades our very descriptions of the world.

It might be objected here that Wittgenstein is not necessary to obtain such a perspective on the political dimension of human activity. This is partly true and partly beside the point. It is true in the sense that such a perspective might just as easily be derived from, say, a certain reading of Hegel and Marx. However, the point here has been in part explicitly to break new ground by teasing the implications of Wittgenstein's later philosophy for issues with which he has hardly been associated (it also compliments Charles Taylor's efforts to link Wittgenstein to Hegel in comparing the relationship between Wittgenstein's objections to the idea of private language and Hegel's critique of the notion of immediate knowledge). But there is yet something specifically Wittgensteinian which has not been explored yet which has very important implications for what is normally termed political science.

It would seem that political science would have to take on a universal character if all parts of the social world are potentially political. Political knowledge seems to encompass everything. However, there are strong reasons for thinking that Wittgenstein's view of rule-following entails the view that we cannot get into a position from wich we can know (in the scientific sense) the social world. To see why this is so will be to articulate, however imperfectly, something which is implicit but unarticulated in Wittgenstein's later philosophy, i.e., the relationship between rule-following and his injunctions about what it is to do philosophy (the form of his philosophizing).

Wittgenstein's remarks that philosophy can only describe practices, that it should not aspire to alter them, have been construed as politically conservative as well as capriciously perverse. This is because they have been read in isolation from his central notion, i.e., to acquire a concept is to follow a rule. It must be added that Wittgenstein himself is not especially illuminating about this connection; so it is not altogether strange that such difficulties with interpretation should arise. Even if what has been alleged concerning the political stipulations of rules and the conceptual link between family resemblances and essential-contestability, there is one point where Wittgenstein and the political analyst must converge. They must both describe practices. It is not often emphasized that, although Wittgenstein insists that description must replace explanation, (1) there is precious little description of practices in his works, and almost none of actual practices and (2) this is probably not accidental given the difficulties inherent in describing our rule-following behavior. This latter point is slowly being recognized by some social scientists. To see why this should be so it will do well to recall what is required of a description of action — something philosophers from Aristotle onwards have unquestioningly assumed we could do.

The first difficulty in describing action arises from the fact that our accounts of action do not have to be correct for them to be orderly and "useful" to us in the sense that they facilitate self-understanding. It is quite possible, for example, for a person to believe that a diet composed completely of grapefruit is healthy and, therefore, to eat only grapefruit for the sake of his health. Such a diet is far from healthy. We can think that we are behaving healthily when we in fact are not. So, the first rule as regards descriptions of behavior is that first person accounts of what someone is doing need not be correct. Paradoxically, they are indispensable; for it is not possible to decide what someone is doing unless we know what he thinks he is doing. A second problem in interpreting action is presented by the fact that there are any number of actions which cannot be understood unless the interpreter has mastered the practice in question. This is because he will not be in a position to recognize what counts as a mistake unless he already knows how to do what he is observing. Moves in chess, calculations, blessings, toasts and the like provide examples of actions that are opaque to individuals who could not in principle be performing them. The interpreter's skill is thus bound to his own experience as much as it is to the intelligibility of what he observes. There are, then, beyond doubt serious difficulties surrounding the interpretation of action, only two of which have been touched upon here. But this is surely no secret except to the naïvest positivist. Wittgenstein's position here is interesting precisely because it is so *radical.*

For Wittgenstein the task of philosophy is to describe; but Wittgenstein does not himself describe much by way of actual behavior. His other account of what it is to do philosophy, i.e., the idea that the task of the philosopher is to assemble reminders, is at once more illuminating of his own work and more consistent with his insistance upon the primacy of rule-following in understanding meaning and action. There are, he insists, things about which it is difficult to remind ourselves. It is precisely this difficulty that obscures the roots of thought in action. One suggestion is that our difficulty here arises because it is possible to follow a rule without being aware that we are doing so in any explicit sense (it is probably not accidental that Wittgenstein gives no examples of the kinds of rules we actually follow but only *evokes* what it is to follow rules except in discussing games like chess). This can happen because we are taught rules as techniques to be applied (as judgments). Along with those techniques we are given examples — or better, we learn those techniques through grasping examples as examples of regularities — which we can fall back upon when there arises some difficulty about how the rule should be applied. He sometimes likens this to using a color-sample to check up on whether a given color is in fact, say, puce. As long as we can rely upon these examples, we know our way around. Our problems arise when we, for whatever reasons, no longer can rely upon them. The result is confusion. The point is that to apply these rules, to use our color sample, we do not have to be able to explain what we are doing. Thus we can have problems explaining ourselves when we are asked to even though we know what we are doing (so maybe Socrates missed the point when he found it a failing in the artisans that they could not give an account of what they did!).

At this point Wittgenstein becomes frustrating indeed; for he only tells us *that* there are things concerning which it is difficult to remind ourselves and *admonishes* us that we should not try to explain why this is so but simply describe our practices (language games). Yet, there is clearly something that requires explanation here. Why are we in this position? Is this a matter of fact or of logical grammar? Wittgenstein is silent beyond a few scattered hints. These hints point again to what is involved in following a rule. To follow a rule we do not have to be able to report upon the technique we have mastered. The point is that it is conceptually impossible to do this in any *rigorous* way. To do so would be a matter of being able to say clearly what it was like *not* to have mastered that technique, which is impossible before mastering it, because we do not know what it is that we do not know (we do not know our way around). After we have mastered the technique (say, simple arithmentic) phenomena are available to us ("the numerable") which were not before. Learning rules has opened up a range of possibilities which were literally unimaginable to us before mastered the technique. But, having mastered the technique, we have only the foggiest idea of what it was like not to master it. In this respect my knowledge of *myself* before I could speak or calculate is not essentially different from my knowledge of another person's color blind state. To understand myself in this situation I must turn to others (as Augustine points out in the *Confessions*). Because this is the case, philosophy cannot be carried on in the manner of a theory; for we cannot be in a position to describe in detail what it is to follow a rule. We can only do this *indirectly*, i.e., remind ourselves of what we had to learn to be able to do what we do. We can form no exact comprehensive picture. So, it is not that we can say nothing whatsoever but that there is a conceptual (grammatical) limit to what we can say. All we can do is to *evoke* our grasp of what it is to follow the rule; for the experience upon which that activity is based is not our own and only indirectly available to us. Nor will anyone else be in a position to do this for me. The bases of our action, the training upon which concept use rests, does not lend itself to scientific analysis. It is not simply that the rules do not form a system, but that we are not in a position to formulate them in any *precise* sense. This is why Wittgenstein *must* explain language through *gestures* such as the unanswered question, the example, the analogy. It further illuminates why all understanding must be historical and also why history must be open-ended.

The point of all this is by no means radical skepticism but, rather, a genuinely modest skepticism. We can clearly formulate pictures of many of our practices. We do have more than a vague grasp of that form of life which was, say, the Italian Renaissance. So, it is not impossible to obtain a perspective on rule-following. What we cannot do is to provide a comprehensive systematic analysis of "the social system" we are enmeshed in. If this is right the very concept of a *science* of action (a social science) is part of an inverted world picture. We cannot ask what the necessary and sufficient conditions for action (behavior to the intellectually unwashed) because there is no action as such only rule-following within the context of specific practices (language games) in terms of which words and actions are woven together. This can be grasped in two ways: contextually, from without, as it were, by assembling reminders in aid of a perspicuous representation of the peculiar "nesting" of practices which constitutes a certain form of life, say, the Italian Renaissance; and, from within, by the experienced individual but in a mode that is tied to his experience in such a way as to be inarticulable apart from it (what Thomas Aquinas called knowledge by co-naturality). What is wholly clear is that the kind of "objective" description demanded by "operational" analysis is simply not available; for on Wittgenstein's account there are transcendental reasons why we cannot observe ourselves.

At this point we may ask how the politics of rule-following relates to "politics" in the mundane sense. Some suggestions have already been made. The one which is perhaps most important is that rule-following produces a kind of regularity in our behavior which limits the ways we can expect to alter our practices. To change society has to be, not simply to change the ruling ideas, the substance of the rules we abide by, but to change our mode of rule-following itself. This latter pretty clearly cannot be decreed from above, since we are not in a position to articulate just what we are doing precisely when we are under the sway of sets of rules. For this reason the Wittgensteinian political analyst ought to be very interested in the flexibility of rule-following behavior in the society he studies. This will clearly have not a little to do with the way in which the practices which constitute that society "nest". It will be of the utmost importance to discern how meaning-constituting practices intersect and determine one-another.

Finally, we may sum up the promise of a Wittgensteinian political philosophy provisionally (and the provisional nature of this enterprize cannot be overstated) as follows: first, it established how constraints are built into those rules through which our very ability to act is constituted; second, the family resemblance character of concepts illuminates how it is at all possible for there to be a conceptual dimension to politics, i.e., how political conflicts are rooted in rule-following; third, the specificity involved in mastering a technique explains why there cannot be a-contextual criteria for rationality as, say, Kant would suggest; fourth, the Wittgensteinian assault upon private language, often derided for its abstruseness, in fact casts light upon how politics is structurally (potentially) present throughout human activity and not merely in one sphere of action; finally, the difficulty in representing our own mastery of techniques to ourselves (not simply the difficulty in answering the myriad questions Wittgenstein puts to us in his later work) indicates that grasping, let alone altering, the rules we follow is radically limited by our very rule following activity. It would be folly to think that the Wittgensteinian perspective on the political here developed is anything other than crude but crude is not the same as unilluminating — nor is it incompatible with a certain promise of further accounting for the ways our conundrums about the political are rooted in entanglements in the rules we follow.

CHAPTER VI

CULTURE, CONTROVERSY AND THE HUMAN STUDIES

The conceptual difficulties which the human studies (humanities, *Gesiteswissenschaften*) confront directly, in sharp contrast to those encountered by the natural sciences in the course of their development, peripherally, as it were, turn upon the problematic character of the very *categories* through which the humanistic areas of research are constituted. In the human studies conundrums, paradoxes and even contradictions arise as soon as we begin to reflect upon their subject matter and not on the frontiers of disciplines as is more typical in the natural sciences. The problems typical of the natural sciences arise as we try to solve problems or answer questions; in the human studies the most pressing problems are typically problems about the nature and form of the inquiry itself. In science disagreement turns upon how we are to answer specific questions; in the human studies it is more a matter of disagreement about what questions to address. Indeed, the most distressing problems that arise in the human studies tend to bear upon the very nature of the subject under investigation. "What is literature?" "What is religion?" "What is society?" "What is it to explain something?" In the human studies these questions are not luxury items as their counterparts in the natural sciences tend to be. When these questions raise their ugly heads the very foundations of disciplines are being called into question.

What is worse the sorts of answers these questions tend to evoke have a way of being annoyingly trivial or even more annoyingly imperialistic. Triviality usually arises when we try to make our answers comprehensive; whereas imperialism results from the desire to establish clarity where there is none with the result achieved is forced and artificial. Little wonder, then, that the common response to the conceptual conundrums surrounding the human studies is to ignore the problem and to get on with specialized research. But this is little more than assuming the obscene posture of the ostrich. Conceptual problems do not disappear. They must be faced head-on in all their bewildering complexity; for to shun them at the outset is to risk compromizing the results of research. All of this is hardly a secret.

Karl-Olof Arnstberg's contribution to this discussion, "The Concept of Culture as a Core Symbol",[1] is distinguished for its author's efforts to come to grips squarely with the problems of the apparent chaos surrounding one particularly crucial concept in the human studies. That means to discuss the lack of clarity and inconsistency surrounding the multiplicity of senses that the term 'culture' has taken on by making and explicit effort to avoid triviality and imperialism. His pragmatic perspective — both in the sense that his primary concerns arise from the *practical* implications of adopting a specific notion of culture at the exclusion of others and the pluralistic outlook so dear to William James — make him sensitive both to how little experts can accomplish by way of defining the problem out of existence and the political character of taking a stance on the question, i.e., the political implications of a specific concept of culture, say, as informing the practices of an ethnological museum. In attempting to sort out the ways in which the term culture functions — principally as referring (1) to "high" culture such as opera etc., (2) as referring to customs or mores, (3) as referring to a sense of quality or good taste and (4) as referring to a set of values (to mention but four of the most important variants) — Arnstberg's efforts are as noteworthy as his problem is pressing.

He rightly concludes his study with the admonition that we have to live with a certain confusion; for all efforts to be yet more precise will in the end make things worse. I do not propose to add anything of substance to those conclusions but to offer a perspective on his pluralism which will tend to illuminate why we need not conceive these conflicts and disagreements as a state of confusion. This I take to be at least one primary function of philosophy with respect to the human studies: to lay bare what Wittgenstein called the "logical grammar" of the concepts that humanists employ in such a way that we come to understand why questions about those concepts tend to be so perplexing. Philosophy's role here is, thus, at once hermeneutic and analytic. Its aim is to articulate the conceptual roots of the sort of problem that Arnstberg has identified. In practice it has more to do with *dissolving* than with resolving quandries, which is precisely what Arnstberg has seen intuitively as being required here.

What philosophy can do in this specific case is to show how the conundrums surrounding the concept of culture are neither abberations nor, for all their disputed character, cause for despair. In fact, 'culture' is a perfect example of what W. B. Gallie has termed an "essentially contested concept".[2] In what is little less than a landmark paper in the area of cultural analysis with that name in 1955 Gallie sought to explain how it was, contrary to the then current account of value disputes, that disagreements about value questions could be at once irresoluable yet rational in nature. In short, he aimed at showing how there could be legitimate conflicts over the meaning of concepts like 'art', 'Christianity' and 'democracy' on the basis of different, but legitimate, criteria for determining just *what* something is (the crucial philosophical question on Aristotle's view it should be remembered). His paper turned on the example of the concept 'art'. Significantly Gallie compared art with the two above mentioned concepts to tease out its essential contestablity. I take Arnstberg's appeal to the similarities surrounding 'culture' and 'democracy' as an implicit recognition of the essentially contested character of 'culture'. Indeed, everything about his elucidation of the problem of culture cries out for an articulation in terms of essential contestability.

What is an essentially contested concept? It is Gallie's merit in developing this cruicial tool for cultural analysis in clearly indicating seven criteria for essential contestability. His point in articulating this notion in the first place was to enable us to distinguish it from radical confusions which result from indiscriminate use of concepts in ordinary street corner arguments. What are those criteria? First, the concept must be *appraisive*. It has to refer to an achievement (e.g. 'work of art' as opposed to '10 1bs.'). Second, it has to be *complex* (e.g. 'democracy' as opposed to 'blue'). Third, the achievement in question has to be the result of the ordering of the parts (e.g. the achievement involved in being 'democratic' involves a particular ordering of the notions of freedom, equality and representation). Fourth, this order will be variously describable (e.g. freedom or equality or representation could be taken to be the prime component in 'democracy'). Fifth, the concept must be "open-textured" (there are new ways of being 'democratic', while there are none for being an isosceles triangle). Sixth, both parties must recognize *some* common authorities (e.g. all democrats recognize John Locke, Jean-Jacques Rousseau, John Stuart Mill as authors of cannonical texts, which have to be given some weight in all discussions of 'democracy'). Seventh, both parties must be committed to further discussion of the matter, i.e., to the idea that it is at least in principle possible to win the debate on the basis of persuasion alone. It should be clear from the very statement of these criteria that they fit the concept of 'culture' perfectly. So, disagreements surrounding it can be legitimate because there are innovative ways of being 'cultural', there are various ways of describing what falls under the concept and there are different well-established traditions according to which 'culture' can be construed differently. Indeed, once the notion of an 'essentially contested concept' becomes available it is hard to conceive culture in any other terms.

William Connolly has advanced Gallie's position in his all-too-little discussed *The Terms of Political Discourse* by identifying essentially contested concepts, not only as the rational basis of disagreement, but also, for that very reason, as the rational basis of politics itself.[3] To be involved in a 'what is it?' debate is, for better for worse *eo ipso* to be in the midst of a political discussion regardless of what specific sphere of human activity is under discussion. Essential contestability, then, is the defining characteristic of politics as a rational activity. Connolly has rightly suggested that the essentially contested character of concepts is tied to what Wittgenstein termed their "family resemblance" character; what falls under a concept is not reducible to any one defining characteristic but rather groups itself around a set of foci (something Wittgenstein explains in the basis of the fact that there are many ways of following a rule). In any case, on the basis of essentially contested concepts we can at once grasp something of the source of our problems with the various ways of construing the concept of culture as well as the political problems attendant upon it which so disturb Arnstberg. The point of recognizing a concept as essentially contested is to see that our opponents in such debates do not *have* to be arguing in bad faith (i.e., that it is possible for them to disagree with us legitimately) even their position is not only different from ours but actually incommensurable with it. This will have made a crucial difference in the attitudes of the competitors. Moreover, it is clearly indicative of the way in which evaluation is intimately linked to description. If I describe abortion as, say, a routine gynecological operation; whereas my opponent describes the same act as killing and unborn person, our respective evaluations of the acts are all but entailed in the act of description. Here we see very clearly just how problems of evaluation are often all but wholly rooted in our descriptions. The important lesson to be learned here is that, *pace* C. L. Stevenson and that ilk, description is a political and a moral act. Indeed, it is the case that who controls description controls evaluation too (if it makes any sense to speak of 'control' in these matters). Furthermore, if our vocabulary is shot through with such concepts, we ought to expect that there are many instances in which the best we can hope for as the outcome of a disagreement is to agree to disagree. This is because in such situations victory cannot be imposed upon the opponent without changing linguistic usage — something more easily tempted than accomplished by *force majeure*. The impact of revealing the essentially contested character of a dispute ought to be to make the debate less vitriolic, more recognizably complex and, above all, recognizably political. Moreover, recognizing essential contestability involves at least the possibility of maintaining communication in the face of significant disagreement. Basically, the possibility of agreeing to disagree makes dialogue possible accross perspectival barriers — it, of course, by no means assures it. It leaves the parties in question on speaking terms and in a position where it becomes clear that it is worthwhile to explore the development of the concepts involved. For either party to 'win' the contest, i.e., really to convince the opponent, this will be absolutely necessary. In short, the need to persuade the opposition at this point should make participants in such debates conscious of their opponent's modes and strategies of reasoning as well as of the development of the concept involved. This illuminantes the role that history can play in debates about the meaning of terms such as 'culture' and 'democracy'. Far from providing a pseudo-solution to problems such as "what is culture?" by providing normative values from the past (something Arnstberg rightly regards as pointless), conceptual history helps us to explain why it is that we find ourselves at our current impasse and, with luck suggests hints at ways beyond it.

Moreover, the contested character of concepts seems to remind us of their teleological character. William Connolly emphasizes that once we realize that we are in a debate over such a concept, we should become no less aware of the importance of C. S. Peirce's trichotomous account of meaning: A symbol *stands* for *something* to *somebody*. 'Theories', conceptual schemes, hermeneutic devices and analogies are always used by specific persons with specific goals. The cultural debates we find ourselves involved in cannot be clarified in the absence of an understanding of the *contexts* in which they are embedded. So, the whole ensemble of practices and traditions in which the participants are rooted must be taken into consideration if we are to understand fully the nature of the contests in which we are embroiled. As a practical suggestion this amounts to the admonition that we are unlikely to prevail in conceptual debate unless we are prepared, contrary to our instincts, to reflect *seriously* upon the kinds of considerations which make the position with which we disagree plausible to our opponents. Only then shall we be in a position to persuade them. In short, we must never forget that in situations where we find ourselves arguing over essentially contested concepts those arguments which strike us as most compelling are precisely those which will strike our opponents as least compelling and vice versa.

In terms of Arnstberg's primary concern this is to suggest that classical humanists (i.e., Latin and Greek scholars) have something to learn from the perspectives of ethnologists and anthropologists as well as social critics, philosophers of culture and even theologians. This is no less true of any of the others in relation to each other, I would hastily add. It is, further to suggest that all of the participants in the debate over the meaning of culture fail to do justice to their own perspective, if they fail to take the perspectives of others into consideration. In the end, the lesson of all of this is that it is at least sometimes possible to turn confusion and acrimony into healthy and constructive discussion. However, this can only happen when we have a reflective grasp of why the situation Arnstberg describes *has* to be the way it is.

This perspective on disagreement (and, ultimately, competition in intellectual life) does not create heaven on earth but it is a start in the right direction inasmuch as it admonishes us to proceed from a realistic grasp of our situation. Consider, for example, how it illuminates Arnstberg' problem with the political nature of the institutionalization of 'culture' in ethnological museums. This problem presents an interesting parallel to one which Gunnar Danbolt has raised with respect art museums.[4] Danbolt has argued that, since the physical surroundings of a painting — think, for example, of a fresco in a Gothic cathedral — is part and parcel of the context against which its meaning must be assessed, to wrench said fresco from said cathedral and install it in a museum is to distort, if not transform, its meaning. This is a thorny issue with many facets to it to say the very least, but there seems to be a prima facie case that the very institution of a museum is to some extent culturally imperialistic by its very nature. *If* this is the case, as it seems to be, the above analysis suggests that this need not be a disasterous defect; rather, it suggests that we should be aware that, like all aspects of human life which have roots in complex traditions, this is precisely what we should expect.

This view that politics has its roots at least in part in rational disputes over meaning has the salutary effect of removing politics from the the exclusive concern of 'politicians' and turning it into a constituent element of normal life. This is important; for, to paraphrase Talleyrand, politics is too important to leave to the politicians (and political 'scientists'). Museums are, indeed, inter alia political institutions, but this is only problematic to the extent that we fail to recognize that fact and/or confuse politics with some sort of 'unclean' practice which human beings must *suffer* rather than with something implied by our very nature as speaking beings. This is, of course, not to suggest that politics is always and exclusively rational. However, as soon as we recognize that it is an inevitable element in human discourse, we put ourselves in a position where we know what to expect, i.e., that it should be a matter of course that such 'political' debates arise.

Thus the notion of an essentially contested concept illuminates many of the conflicts in which we find ourselves enmeshed precisely because it de-mystifies them. We move from a situation in which confusion and frustration are — or at least can be — replaced by the recognition that we are using the same term accross incommensurable perspectives and for that reason our debate must be political. But that very recognition brings with it the possibility of replacing polemic with persuasion as befits pluralistic rationalism. I suggest that the human studies will only come of age when their practitioners realize that such problems about the politics of interpretation lie at their very core, and, consequently, must be embraced rather than shunned.

NOTES

[1] Karl-Olof Arnstberg, "The Concept of Culture as a Core Symbol", *Methodological Questions* no. 2 (Stockholm, 1985). Professor Arnstberg himself finds these remarks illuminating. I am grateful for the opportunity he graciously provided to discuss my reformulation of his views in a seminar at the Ethnological Museum in Stockholm in May of 1986.

[2] William Gallie, "Essentially Contested Concepts", *Proceedings of the Aristotelian Society* 56 (London, 1955—56); reprinted in *The Importance of Language*, ed. Max Black (Engelwood Cliffs, 1962), pp. 121—46.

[3] William Connolly, *The Terms of Political Discourse* (2nd. ed.; Princeton, 1982).

[4] I have greatly benefited from discussions of contextualism in art history and the political nature of museums with Professor Danbolt in Bergen (Norway). See also his "Esthetic Theory and Practice in Art History", *Contemporary Esthetics in Scandinavia* ed. Aagaard-Mogensen and Hermergén (Lund, 1980).

CHAPTER VII

THE POLITICS OF CONCILIATION

Forecasting the future of anything whatever is heady business — all the more so, if the subject in question is philosophy. If we need reminding of that, we might just consider for a moment how the future of philosophy looked to Hegel in 1806, to Marx and Feuerbach in the early 1840s, to Russell and Moore at the turn of the century or to Ebner and Jaspers just after World War One. Perhaps the best way to approach this difficult subject is to consider the attitudes of some prominent thinkers to the future generally.

Consider John Dewey and Otto Neurath. For both of the fathers of philosophical Naturalism the future is something that is radically *open*. Anything can be accomplished once we eliminate demon metaphysics and finally rid ourselves of the notion that philosophy can provide theoretical foundations for knowledge. Both Dewey and Neurath were confident that once the whole of the intellectual community threw itself whole-heartedly behind natural science all serious obstacles to remaking the world along rational lines would be removed.

Martin Heidegger provides us with a second view of the future, one which takes just the opposite view of our ability to shape the future. For Heidegger we are at a turning point in the history of rationality. Today the whole scaffolding which makes rational belief possible, i.e., canons of evidence, principles of inference, etc., has collapsed. On this view, the Naturalist's view of the reconstruction of society is the hight of folly, for there is nothing to build a future upon. So, efforts to transform society are doomed to be nothing more than arbitrary and, in the end, self-destructive. On this view, which is more deserving of attention than it has been afforded by Anglo-American philosophers (which is not yet a complete endorsement), the Owl of Minerva doesn't even get off of the ground but only seems to do so, frenetically flapping its wings in its efforts to eliminate metaphysics, naturalize psychology, conceive social relations dialectically and historically or deconstruct texts.

Karl Marx provides a third way of looking at the future. As a solid Hegelian Marx, too, was in principle opposed to pronouncing upon the course of future developments, but, when circumstances warranted, for example, when the German Social Domocrats produced their Gotha Program, he was not above making some general remarks about what we should *not* expect the future to be like. We should not, he suggests, expect the future to be dramatically different from the present. To the extent that we get a glimpse of the future at all, we do so by learning to see how it exists seminally in the present. This is indeed heady business. To the extent that this is at all possible, it is only possible on the basis of an accurate assessment of the situation in which we find ourselves. History, as another, very different Hegelian, R. G. Collingwood, knew, is vital to this understanding — not as an account of the past, but *as the source of our understanding of the dynamics of change*. It is this position which informs my remarks

Pace Dewey and Neurath, metaphysics and foundationalism are more or less a thing of the past. They may persist but they are no longer the formidable forces that they once were. Nevertheless, we seem no closer to the scientific-democratic millenium than we were fifty years ago. Scientists have not turned out to be the vanguard of a new society à la H. G. Wells; nor has science turned out to be immune to ideology as the French epistemologists have shown. *Pace* Heidegger, the world around us is not completely chaotic intellectually; communication may be sporadic and difficult, but it, nevertheless, still takes place. To be sure, our traditions are considerably less reliable guides than we once thought, but they are not wholly bankrupt. Some of our projects for reform, even at the international level, do manage to get off the ground. It seems less that we are wholly unable to control our development, than that we have at best a flawed and imperfect understanding of cultural dynamics. If this is right, it seems safe to assert that the future, however different it may turn out to be than the present, is unlikely to *look* much different from the present. Therefore, the task at hand with regard to the future of philosophy is to ask how the future is present right now. This effort to see the future in the present has two aspects: determining our actual situation with its inherent dynamics (i.e. rule-following patterns) and identifying those extant ways of doing philosophy which already manifest co-operation rather than confrontation between philosophical traditions.

In general, I am in agreement with Richard Bernstein when he asserts that the distinction between analytic philosophy and phenomenology, hermeneutics etc. is increasingly less meaningful — *philosophically*. But, if that is the case, why do we continue to be troubled by that distinction to the point of convening symposia to disucss it? If I am right, it is because the present and the future of philosophy is not merely a matter of the substance of philosophizing alone but also of the sociology and politics in which contemporary philosophizing is embedded. It is these latter that I propose to disucss with a view to illuminating the former. Limitations of space prevent me from arguing my case in detail, so much of what I shall say has to be in the form of boldfaced assertion. In defense I would like to point out that I have argued some of these points elsewhere and am currently preparing yet other detailed accounts of the sorts of things I shall be asserting here.[1]

Philosophers are sometimes good sociologists but they are nearly always terrible sociologists of philosophy — it should not pass unremarked that professional sociologists would appear to consider philosophy too insignificant to merit study! There is probably a moral here. One case in point with respect to the low caliber of philosophical sociology of philosophy is William James, whose tough/tender-minded distinction can at best be described as feeble sociology (James, of course, did not see himself in exactly the role of a sociologist but it is not absurd to to suggest that this was really what he was after in proposing his classification). More recently we have the case of Richard Rorty's contemptuous attitude to the sociology of philosophy. With Villiers de l'Isle Adam's Axel he seems to say: as for the sociology, our servants can do that for us.[2] Rorty has chosen Thomas Kuhn as his servant in this case. The former's classification of philosophers is based on the latter's distinction between normal and revolutionary science. However, this seems to be a dubious procedure. For starters, Toulmin has questioned the adequacy of that distinction in the sociology of science.[3] If it is dubious in its home setting, it seems less than astute to transfer it to another without making the slightest effort to see whether it is appropriate in its new surroundings (I am not suggesting that philosophers have not seen themselves in terms of the categories Rorty employs to classify them but, rather, questioning just how important it is that they have done so). One kind of problem Rorty glosses over in adopting that distinction is the problem of false consciousness. The first thing you learn as a social scientist is that people (philosophers included!) routinely describe themselves inadequately in all sincerity. Indeed, this is perhaps the most important meaning of the term *ideology*. This happens in philosophy too. For example, a French philosopher, who had ironically identified himself as wholly French in his efforts to explain why he was *not*

French, recently made a plea on behalf of analytic philosophy, which, in fact, was not a treatise *in* analytic philosophy but almost wholly Nietzschean in tone.[4] Such examples could easily be multiplied. The point is that Rorty fails entirely to take this sort of thing into consideration.

While on the subject of Rorty — and in light of the fact that several of our speakers have drawn attention to the fact that philosophers have tended to see the history of philosophy in 'whiggish' terms, i.e., as leading up their own enlightened views, I feel obligated to point out that Rorty sagely insists that whiggishness is not wholly avoidable in the historiography of philosophy inasmuch as that history is written by philosophers themselves, who seem unable to see the matter differently (this implies, contrary to almost all philosophers' views of the matter, that there is a warrant — perhaps even a demand — for histories of philosophy which are *not* written by philosophers). We do well to remind ourselves as philosophers of something that historians are conscious of, namely, that there is *no history as such* there are only histories whose teleology derives from various perspectives on specific sets of developments, which are often very different from one another. In the case of the history of philosophy, what constitutes this point of view is precisely what we take to be the most crucial developments in the history of philosophy. Because our histories turn upon what we take to be philosophically crucial, they are bound conceptually to our philosophical positions. This is important because it implies that, if we cannot establish that philosophy rests upon logical foundations, neither can we establish that it rests upon historical foundations, i.e., which legitimate its questions and techniques. If someone insists that what counts as legitimate philosophing rests upon historical precedent, we should ask "according to whose history?".[5] But let us return to the question of how the sociology of philosophy might illuminate its present and provide us with some inkling of the future.

Since there is at the moment precious little which counts as sociology of philosophy, we must first invent such a study in order to see its implications. A sociology of philosophy must concentrate upon the ways in which, for example, the institutional setting in which philosophy is practiced determines, not simply the substance of philosophical positions, but also what tactics and strategies will count as permissible modes of philosophizing. Thus we must learn to ask, for example, what kinds of correlations exist between the brand of philosophy which emerges when philosophy is taught in `tutorials with two or three students, in large lectures, in seminars or in classes in which there is a combination of lecturing and discussion. It seems hardly accidental that a brand of analytic philosophy which construes analysis as a pure technique, (i.e., divorced from science), whether horseshoe or softshoe (to use the favored expression of Professor Church), has tended to thrive in the first sort of setting; whereas the second has been more conducive to the criticism of whole philosophical projects and world views rather than of arguments.

The sociology of philosophy must also inquire into the relationship between the development of ideas and the shape of philosophical careers. Attitudes to what counts as acceptable in philosophy are hardly separable from the kinds of considerations which advance a career. No one in the field would deny that the tenure system in America or the habilitation system in Central Europe affect respective the substance of philosophing but there has been precious little discussion of precisely *how* that happens. Moreover, the role of charismatic figures in the field is a constant topic of discussion among philosophers in the Faculty Club, but it is hardly ever mentioned in scholarly discussions of how philosophy develops. In all of this I am not claiming to say anything arcane or particularly profound but simply pointing out the ways in which factors of which we are acutely aware ought to function in the ways we publically discuss the development of our field and profession; for they have clearly had a bearing upon the relationship between the traditions, and shall also have a bearing upon any future rapprochement between them.

One extremely important task of a sociology of philosophy will be to provide us with an accurate description of the way philosophers are in fact divided from one another, i.e., on the basis of the ways in which they actually *do* philosophy, as opposed to what they say about what they do. Here it is at least possible to make some suggestions about where to begin. Questions of the following sort seem crucial.

(1) Is the given philosophical position pro- or anti-science?

(2) Is the mode of philosophizing centered upon constructing and analyzing arguments or does it aim at insights which do not permit themselves to be formulated in terms of premisses and conclusions?

(3) Does that mode of philosophizing seek to produce a theoretical foundation for knowledge, i.e., some sort of discourse about the necessary and sufficient conditions for discourse?

(4) Is the aim of philosophy to produce theories?
(5) Does human action 'describe itself' in some sense or other or does it always require interpretation?

The model this questionnaire suggests is, of course, almost grotesquely oversimplified. There are more possibilities than it admits of. Clearly it is possible, for example, for a position to be sympathetic or unsympathetic to science generally without actually being pro- or anti-scientific. Moreover, specific cases of scientific theories such as the alleged sociobiological reduction of ethics complicate matters in the sense that even someone who is otherwise committed to the idea that philosophy is continuous with science might find the aspirations of a particular research program such as this objectionable. Similarly the question of the role of argument cannot be answered till we specify what we mean by argument (the question "what is an argument?" is one which philosophers and historians of ideas have curiously neglected). There are comparable problems with the idea of "theory" and "foundation". So, a sophisticated reading of these questions would yield a spectrum of opinion rather than two neatly divided camps — which is as it should be; for neat divisions in twain almost inevitably lead to rejection of one pair for the other. Nevertheless, even raising these questions in a naive manner yields interesting results immediately; for the distinction between 'analytical' and 'Continental' philosophers begins to dissolve at once. Husserl, for example, ends up on a different side of the fence from Heidegger; while Wittgenstein is similarly separated from Russell, Carnap and Searle. The point of all of this is to emphasize systematically that there are in fact 'analytic' philosophers who have more in common with 'Continental' philosophers than they do with their fellow 'analytic' philosophers and vice versa. But this is to say that the distinction is at least in part *political*, which brings us to the crux of the issue: *it is characteristic of philosophical quarrels in this century that they are shot through with politics in a way that philosophers themselves have been reluctant to recognize.*

When I speak of politics here I am referring to a kind of politics which is immanent in the attitudes of philosophers to their subject, not to party politics, although that too can sometime play a bigger role than we are normally prepared to recognize. When I speak of the politics of philosophy, I am referring to the aspirations of particular philosophical programs with respect to one another — and, therefore, to the structure of philosophical *disagreements*. The aspirations of philosophical programs explain how politics enters into philosophy; determining the structure of philosophical disagreements can help ùs to deal with this immanent political element in our philosophizing sanely. The sorts of disagreements I am referring to are as old as philosophy itself. Think only of the quarrels between Aristotelians and Epicureans over the the primacy of justice over friendship in the good life. My concern is with more recent aspects of these political struggles within philosophy.

All of the major movements on twentieth century thought have been embedded in images of what society ought to be like in ways which exactly parallel the differences between the Aristotelians and Epicureans of yore over the respective merits of the private and the public spheres of life. Thus Viennese positivism was part of the intellectual weaponry of Viennese liberal democracy. It was not altogether accidental that the Vienna Circle became identified with socialism despite the a-political, or even conservative, politics of some of its members. Hermeneutics, on the other hand, (at least in its predominant German version) took this *anti-authoritarian* attitude to be *anti-traditional*, and, therefore, uprooting, as, indeed, it was. The political element in each of these programs can be identified in terms of that element in human affairs that each would *suppress*. Indeed, each of these philosophical movements advocates an ideal of intellectual hygiene, which is to be cultivated in part by a *strict avoidance* of certain sorts of thinking. I have termed these modes of cultivating a specific set of personality traits *spiritual practices* on the analogy with religious spiritual exercises such as those of Thomas à Kempis, John of the Cross or Ignatius Loyola.[6] They are most easily defined in terms of what they admonish us to put out of our minds. Thus, Neurath would have us cultivate the scientific conception of the world by purging ourselves of all interest in ethics inasmuch as it (paradoxically) involves cultivation of "character", reflection about the meaning of life (*Gesinnungsethik*), etc. We should shut every view of ethics but Benthamite Utilitarianism out of our thoughts. For Hans-Georg Gadamer on the other hand, we should systematically forget about everything scientific and technological for the sake of cultivating the (principally, Greek) classics collectively. Now, I take it that sensible people recognize that we would be hard put to dispense with utility in ethics and that an education which wholly omitted the classics is radically and deplorably deficient; however, sensible people also recognize that it is the height of folly to disregard questions about how we should form our character or to consider science and technology as the Devil's work, (even Heidegger believes this!). To do so is not only capricious; it is *political* in the sense that I have used the term. This is exactly where politics enters twentieth century philosophizing. But, why, we may ask, did the political element in twentieth century philosophizing become so important, i.e., to the point of obscuring important philosophical similarities between analysts and phenomenologists? The answer; I think, also has to do with politics; this time in a more easily recognizable sense. *It bears at least in part upon the ways philosophers have connected modes of philosophizing with military causes in this century's major wars.*

Consider the following assertion penned in 1950:

> Neither idealistic phenomenology nor existentialism could long survive on the American scene, assuming it could get a sufficient foothold. With the systematic arguments for idealism exploded in the First World War, the defenders of idealism could only argue from the incompleteness of the sciences.[7]

It is not the statement of a rabid logical positivist but that of perhaps America's leading phenomenologist of the day, Marvin Farber. My question is, "how is it at all possible for such a statement to fall from the pen of such a distinguished thinker?". Why would anyone within the philosophical community assert that arguments are refuted on battlegrounds? Would we ever want to say, by parity of reasoning, that arguments are established there? I doubt it. Strange though it may seem, the answer is because in 1914 philosophy went to war when the great powers did. With the passage of time we have forgotten this crucial datum.

When I say that philosophy went to war, I mean that philosophers interpreted World War One as a conflict of cultural ideals and, therefore, as a counterpart to philosophical antagonisms. In order to illustrate this point I want briefly to consider two cases in point which indicate the ways in which major philosophers viewed the War in books written in 1915. The philosophers are John Dewey and Max Scheler; the books are Dewey's *German Politics and Philosophy* and Scheler's *The Genius of War and the German War.*[8] Dewey needs no introduction; however, a word about the stature of Max Scheler is necessary.

Scheler was the most important of Husserl's early disciples. His standing in the phenomenological community is attested to by virtue of the fact that the young Martin Heidegger dedicated his first book, *Kant and the Problem of Metaphysics*, to Scheler. Scheler left his mark upon phenomenology *inter alia* in the form of a certain conservative political orientation, which was by no means Husserl's, and, which, would pave the way for, say, a Gadamer's classicism. Moreover, it was Scheler, who first suggested that there could be a Christian resoponse to Nietzsche, and, thereby, paved the way for Kierkegaard's entry into twentieth century philosophy (Scheler's pupil, Theodor Haecker, provided the Germans with some of the first reliable translations of the Dane's texts, including *An literaere Anmeldelse* or *The Present Age*). Thus, Scheler is largely responsible for one of the huge differences between Continental and Anglo-American thought: philosophical concern with Christianity (Scheler was the first to recognize the philosophical significance of the works of Cardinal Newman).

Both Dewey's and Scheler's books were replete with clichés. For Dewey the distinction between the noumenal and the phenomenal worlds in Kant's ethics is taken to typify a German attitude to the sphere of action as old as Luther. It is a view, which, Dewey alleged, predisposes Germans to be more concerned with the Ideal of freedom than with its institutionalization. Thus, Dewey could establish a link between Idealist ethics and German barbarism in the trenches. His main evidence for such a direct link comes from the speeches of General von Bernhardi, who was in the habit of citing Kant when exhorting his troops. Of course, German philosophy in the person of Rudolf Eucken also supplied grist for Dewey's anti-dualistic mill. If his efforts to identify Idealism with "brutalism" were clichéd, it was prominent Germans who supplied Dewey with the clichés.

Curiously, Scheler's efforts to convince Germans that (1) the War was a noble undertaking and (2) one that was really directed against the genuine enemy of the German spirit, England, not France — at this time there was still a widespread strong feeling (which Hitler shared till 1940) that the English, who were ruled, after all, by the House of Hannover, were in fact fellow Teutons and, therefore, natural allies in the battle of Germanic "culture" against French "civilization" — was also conceived as a battle against the double standard. Scheler's assault took the form of an attack upon what he termed English "cant". Thus, he points out that the English support for the underdog has in fact been a method of justifying English interests more than it has a defense of the weak. In plain terms, Scheler argues, English public life is imbued with hypocritical euphemisms, of which he supplies a lengthy list, which are in fact defining of English values. The whole argument turns upon representing England as the epitome of mechanistic *Gesellschaft* (i.e., modern, industrial) values, which threatened traditional German *Gemeinschaft* (i.e., traditional, pre-industrial, small town) values. In Sombart's terms, which Scheler employed approvingly, a nation of heros (*Helden*) was being threatened by a nation of traders (*Händler*). All you had to do was to read their representative philosophers, Hume, Smith and Spencer to see this. If all of this reeks of cliché even more than Dewey does, it is important to emphasize how many of these chichés were drawn from contemporary English letters. In fact, Scheler's account of English mores is documented from two sources which Germans to this day regard as infallible: Oscar Wilde and George Bernard Shaw.

My point is that these clichés — and the wartime propaganda in which they were embedded — as much as substantive philosophical considerations that moulded our attitudes to one another. We do well to bear in mind Marx's admonition that all history, the history of philosophy included, is written by the winners. Thus, Dewey's clichés have way of sounding somehow less clichéd to us than Scheler's. In a similar vein, it is hardly a secret that Russell and Heidegger's political actions have had as much to do with their reputations as the substantive merits of their positions. In short, with the wisdom of hindsight and from a clichéd perspective, analytic philosophy won World War Two (having already won World War One) and kept the world safe for democracy at the cost of driving the wedge between the traditions yet deeper.

There is a great irony in all this; for the roots of English analytic philosophy are as much in British Idealism as they are in Frege.[10] But British Idealism was thoroughly imbued with German thought. It is well-documented that the 'revolution' Russell and Moore saw themselves initiating at Cambridge at the turn of the century was carried through with weaponry forged accross the Rhine. From a strictly philosophical point of view this is ironic indeed; yet, from the perspective of cultural politics it is clear that said 'revolution' was but one more Edwardian effort to throw off the moral shackles of Victorianism. What that most Edwardian of Edwardians, G. K. Chesterton, said of the nineteenth century could just as well have been expressed by the philosophical parricides, Russell and Moore in rejecting, not only McTaggart and Bradley, but also Carlyle and Ruskin:

> There is no more remarkable psychological element in history than the way in which a period can suddenly become unintelligible. To the early Victorian period we have for the moment lost the key. The Crystal Palace is the temple of a forgotten creed . . .

[there is] a nineteeth century atmosphere which is so difficult to describe that we can only convey it by a sort of paradox . . .[11]

I suggest that this extraordinary account of the distance between the generations at the turn of the century was constituted by the way Chesterton's generation wanted to see itself. To understand the intensity with which Russell and Moore rejected Idealism, it is necessary to look beyond the texts in which they developed their arguments into the aspirations of a whole generation to liberate itself from the burden of history, tradition — and *style* (there is a lesson for the historian of philosophy in Proust too!) — for the sake of forging its own identity.

In Germany at the same time there is plenty of evidence that analogous social forces were at work behind the scenes shaping the course of philosophical politics. I refer to the debates over the merits of industrialization, democracy and secularization. There is no doubt that German sociology (Tönnies, Sombart, Weber, Simmel — and Scheler) cut its teeth on those complex and bitter debates.[12] There is still less reason to doubt that they scarred philosophy, which developed hand in hand with sociology in Germany at that time. So, it is little wonder that Scheler wanted to cast the War as an ideological struggle between two social ideals; for England was clearly what Germany would become, if Germany continued on the path to industrialization, democracy and secularization in the eyes of these thinkers.

We should not let these similarities obscure one important difference between the ways in which philosophy was embedded in social conflicts in England and Germany respectively; for this difference has much to do with determining the attitudes of German and Anglo-American philosophers to one another. Russell and Moore claimed to have undermined idealism *from within*, i.e., by criticizing its central *arguments*. Once having done so, the attitude of the parricides to their vanquished antagonists was one of disdainful dismissal. Russell, Moore and Co. went about their business as though the Idealists never existed except in a remote period of philosophical *Urdummheit*. Their German counterparts, like German social critics generally, relied upon developing more comprehensive perspectives on whatever subject was in question, which allowed them to "turn the tables" on antagonists, i.e., to show that somehow their opponent's views proved exactly the opposite of what they claimed. In short, radical English philosophers were at home in bourgeois society and felt no compulsion to question that social order; whereas radical German philosophers were not, and, consequently, were compelled to question, not only the views of their opponents, but their very conception of the role of philosophy in intellectual life and in society generally. Echos of these tactics, which were as much recipes for changing the world (in the German setting, often in the name of 'tradition') as for understanding it, reverberate throughout the politics of contemporary American philosophy as we argue over what legitimately counts as philosophy.

I have suggested that the political situation that contemporary American philosophy finds itself in is by no means hopelessly condemned to internal strife. We can live with the political divisions which have exaggerated philosophical differences almost beyond recognition to the extent that we can recognize the political origins of these struggles. This is not especially easy; for it involves a certain amount of good will, which, sadly, is not always forthcoming. However, given a modicum of good will, we do have a conceptual strategy for coming to grips with the sort of situation in which we find ourselves. I am referring to William Gallie's program for distinguishing rationally based disagreements from irrationaly based ones on the basis of the "essential contestablitiy" of the concepts that constitute them.[13] Gallie has brilliantly articulated criteria for distinguishing legitimate disagreements from capricious stubbornness. The basic idea is that highly complex concepts like democracy, art or Christianity can legitimately taken on conflicting senses in the course of historical development. Take the case of Quine's allegation that two kinds of people study philosophy: those interested in philosophy and those interested in the history of philosophy. It is clearly rooted in a long tradition of empiricist thinking, but, nevertheless, remains highly disconcerting to a huge segment of American philosophers. His denial that the very enterprize in which I am now engaged is a philosophical one is one which can be disputed on the basis of other aspects of what has traditionally been termed philosophy than the ones dearest to him. His opponents rightly charge him with attempting a political manoeuvre in so constricting the use of the term. For many philosophers, 'analytic' as well as 'Continental', questions about the nature of philosophy philosophical activity are a very important part of philosophy itself. To be sure, there are a host of problems here, but the point is that they are *legitimate* problems. How are we to deal with them?

First off, there is no guarantee that we can *solve* them. This is what it means to assert that they are *political* problems in the sense that I have been using the term.[14] Perhaps the best we can do when confronted with such a situation is to agree to disagree. Admittedly far from an ideal solution, this at least leaves us on speaking terms — Rorty's cultural conversation can proceed albeit sporadically. But how are we to put ourselves into that position? Gallie suggests that we establish the legitimacy of our respective positions by indicating just how they emerge from different traditions of interpretation, which, nevertheless, recognize common authorities (i.e., as all democrats recognize Locke, all art critics Rembrandt and all Christians St. Paul). If it is worth establishing the correctness of our view, on this account, we are obliged to continue the conversation with the hope that we may eventually convince our opponent, however difficult that may prove. All of this together constitutes rational disagreement in Gallie's terms and is the very distinguishing characteristic of the political *agōn* according to William Connolly, who has done most to explore the implications of Gallie's work for our understanding of politics. I would add that it implies that, if we really want to establish the superiority of our position the situation requires that we abandon the adversarial posture for one in which we attempt to *persuade* our opponents. Proof and refutation have to give way to hermeneutics and the history of ideas, *if we are to do justice to our own position.* In this spirit, I believe, we could enter upon as interesting and important dialogue on, say, the relative merits of considering the philosophy of language as part of the philosophy of mind as opposed to the philosophy of culture. But, in all of this, what is there of the future — or a possible future — of a more cordial relationship between 'Continental' and 'analytic' philosophy than that which is currently extant?

I want to consider three examples of profitable cross-fertilization between the traditions, which can yield clues and models of how we may proceed. The first is an all-too-little-known historical case; the second and third are contemporary.

My first example of significant effort to bridge the chasm between analytic philosophy and phenomenology is the case of the French mathematician, philosopher, mystic and hero of the Resistance, Jean Cavaillès. Cavaillès proceeded from the interesting question, "how is philosophy present *in* mathematics?" rather than seeking a philosophy *of* mathematics. Inspired by reflections upon Kant's views about canon and organon, Cavaillès sought to understand the development of mathematics as a *unity* of the 'internal' history of mathematical problems and the 'external' history of the problems of mathematicians. His resolution to this dichotomy was a wholly immanent, non-foundational *philosophie mathématique* (the title of his main, notorioulsy obscure, work), as opposed to a philosophy of mathematics.[15] This he developed principally out of Husserl, but he was at the same time indebted to Wittgenstein's anti-foundationalist, pragmatic conception of logic in the *Tractatus*, whose ethically inspired mysticism the pious Protestant Cavaillès shared. The full significance of Cavaillès as a philosopher — and as a bridge between the traditions — awaits a long-overdue in-depth study of his thought. However, this much is certain: we ignore him on both counts at out peril.

My second example also bears upon the formal sciences as well as Husserl's work on logic. I refer to Barry Smith's Austro-German Seminar in the United Kingdom. The members of this group have taken it upon themselves, *inter alia*, to rehabilitate Husserl's much-maligned views of that subject. They want to establish that Frege did not have the last word in their famous interchange, that Husserl's mereology, or theory of parts and wholes in the third *Logical Investigation* profoundly illuminates central issues in the philosophy of logic, territory normally taken to be the exclusive property of analytic philosophers.[16] The Austro -German Seminar remains true to the spirit of Husserl both with respect to the range of its concerns and its mode of operation. Like Husserl, the members of the Seminar are deeply concerned with the broader cultural context in which science (in the European sense) is carried on. Thus, their meetings are as likely to address problems relating to, say, political economy, literature or cultural history as they are, say, modal logic. Moreover, true to Husserl's (but also Neurath's) intention, its members see themselves as involved in a co-operative venture. Consequently, they often attend international symposia *en bloc*, i.e., several members will read internally related papers in the same session. Their over-all aim has been to provide a 'phenomenological' solution to a set of 'analytic' problems. In the course of doing so they have been particularly successful in their explorations of the common origins of the traditions in half-forgotten figures such as the much-maligned Meinong, Stumpf and Ehrenfels. Interestingly, they display virtually no concern for the Schelerian or Heideggerian modes of doing phenomenology, probably because these figures are on the opposite side of the science-argument-foundations-theory constellation I spoke of earlier.

My third paradigm of cross fertilization between the traditions has arisen in the development of a peculiarly Norwegian philosophy. Under the aegis of its founding fathers Hans Skjervheim and Jacob Meløe, Norwegian *praxeology*, as it designates itself, has sought to illuminate human action through a conceptual analysis that takes its perspective from Continental philosophy (Scheler, Cassirer and Heidegger) but sees this perspective as requiring a more rigorous articulation than it received at the hands of its originators. Thus, they take one strand of Continental philosophy as profoundly complimented, indeed, *completed*, by a certain strain of analytic philosophy (Wittgenstein, Austin, Anscombe).[17] Like Cavaillès and the Austro-German Seminar, deep concern for substantive problems has moved the Norwegian praxeologists to ignore philosophical party lines. Moreover, confrontations play no role in Norwegian praxeology; rather, the praxeologists are wholly concerned with establishing just how philosophy can help to unravel conceptual knots as they are encountered in art history, social science and everyday life. They are philosophers who are committed to the notion that philosophy must be genuinely good for something in the sense that it must enable real people with real problems to handle those problems more satisfactorily than they otherwise might. In doing so their particular focus is upon what the agent must necessarily know to be the agent that he is.

Perhaps the most impressive results that this hermeneutic of practice has yielded have been in Bergen in the much-neglected area of aesthetics. Here too cooperation has played a crucial role. Close cooperation between philosophers Kjell S. Johannessen and Tore Nordenstam with art historians Gunnar Danbolt and Siri Meyer — the former inspired by Wittgenstein's notion of what it is to follow a rule.[18] the latter dissatisfied with the standard hermeneutic alternatives in their field, i.e., between narrow, technical stylistic analysis and broad analysis of the ways in which a work of art reflects a *Weltanschauung* — conceived a radically new and enlightening understanding of the notion of context. On this account the art historian begins his analysis by asking for what specific *use* a work of art is produced. This amounts to asking about the *physical context* for which it was produced. This mode of questioning becomes enlightening as the art historian comes to ask what kinds of presuppositions must be built into a work of art in that physical context. This locates the work of art, not simply in a culture, but in a specific place — and a-fortiori a specific time — in that culture, which in turn becomes the avenue of access for determining why the form of its execution has been what it has been. One focusses, then, upon the *competences* required to produce art of *this* sort in *these* circumstances. Space does not permit me to go into sufficient detail to do justice to the achievements this approach has yielded but I do want to mention one particularly impressive accomplishment. This is a wholly fresh look at Leonardo's "Last Supper". Gunnar Danbolt was able to identify an aspect of this oft-analysed picture, which had hitherto eluded art historians, by looking at it from this point of view. This is the fact that it is situated in a monastic refectory, but more than that, so situated that the monks' meal was symbolically one with that of Christ and the Apostles.[19] The program upon which Danbolt's work is based combines a sensitivity to hermeneutic problems with an acute sense of Wittgensteinian contextualism in a way that can be instructive to anyone interested in the rapprochement of the traditions. It is in such efforts that I take the possibilities for a more peaceful, more fruitful future to lie.

Let me close by recapitulating briefly my main points. It is not the substantive merits of their positions which divide philosophers but the heritage of their philosophical politics, which philosophers bear all-too unconsciously. Given a certain amount of good will, it is possible to discuss our differences rationally — possible, but by no means easy; for that entails substituting persuasion, patience and the will to listen for the cut-and-thrust, winner-take-all adversarial posture that philosophers have traditionally assumed towards one another. Common concern for substantive issues in other disciplines and *perhaps* common interest in common origins (for example in the thought of William James) *might* prove fruitful starting points. We can learn to live with the politics of philosophy, perhaps even to thrive with it, if we *want* to. We face a decision.

NOTES

[1] See my "Ebner und Popper als Denker" *Akten des internationalen Ferdinand Ebner Symposium*, ed. Walter Methlagl (Salzburg, 1985), and "Wittgenstein: An Austrian Enigma", *Austrian Philosophy*, ed. J. C. Nyíri (Munich, 1981), pp. 75—89.

[2] For the details of my critique of Rorty's views about the future of philosophy see above, pp. 80—92. My criticisms are directed at the views he expresses in *Philosophy*

and the Mirror of Nature (Princeton, 1979) not those in *The Consequences of Pragmatism* (Minneapolis, 1982), whose Introduction and Conclusion compliment the views I have developed here, xiii-xlvii; pp. 211—30.

[3] Stephen Toulmin, "Does the Distinction between Normal and Revolutionary Science Hold Water?" *Criticism and the Growth of Knowledge*, ed. Imre Lakatos and Alan Musgrave (Cambridge, 1970), pp. 39—48.

[4] Jacques Bouvresse, *Le philosophe chez les autophages*, (Paris, 1984).

[5] John Passmore has brilliantly developed the spectrum of opinion concerning the relations between philosophy and history in his contribution to *Scientific Explanation and Understanding*, ed. Nicholas Rescher, (London and Boston, 1983), 83—105; cf. Passmore's "Why Philosophy of Science", *Science Under Scrutiny*, ed. R. W. Home (Dordrecht, 1983), pp. 5—30.

[6] I have discussed the concept of spiritual practice in "Discussing Technology: Breaking the Ground", *Is the Computer a Tool*?, below and "Style and Idea in the Later Heidegger", above.

[7] Marvin Farber, "Descriptive Philosophy and Human Existence", *Philosophic Thought in France and the United States*, ed. Marvin Farber (Buffalo, 1950), p. 430.

[8] John Dewey, *German Politics and Philosophy* (New York, 1915); Max Scheler, *Der Genius des Krieges und der deutsche Krieg, Gesammelte Werke IV: Politische-pedagogische Schriften* (Bern & Munich, 1982).

[9] On Haecker see my, "The Philosophy of Inwardness: Haecker, Kierkegaard and the Brenner", *International Kierkegaard Commentary*, Vol. **I**: *The Two Ages*, ed. Robert Perkins (Mobile: forthcoming).

[10] John Passmore, *100 Years of Philosophy*, (London, 1957), p. 471 *et passim*.

[11] G. K. Chesterton, *G. F. Watts* (New York, 1902), pp. 10—2.

[12] For important studies of the relationship between socio-economic problems and intellectual life in Wilhelmine Germany see Arthur Mitzman, *The Iron Cage* (New York, 1970) and Kenneth Barkin, *The Controversy Over German Industrialization* (Chicago, 1969).

[13] William Gallie, "Essentially Contested Concepts", *The Importance of Language*, ed. Max Black (Engelwood Cliffs, 1962), p. 121—46.

[14] William Connolly, *The Terms of Political Discourse* (Lexington, Mass., 1974), p. 181.

[15] Jean Cavaillès, *Philosophie mathématique* (Paris, 1969). I am indebted to Professor Santiago Ramirez for information about Cavaillès; Ramirez is currently preparing a definitive study of Cavaillès. He presented a preliminary report of his research to the Boston Colloquium for the Philosophy and History of Science in October 1984 under the title "An Alternative for the Philosophy of Mathematics", *Boston Studies in the Philosophy of Science*, forthcoming.

[16] Barry Smith, *Parts and Moments* (Munich, 1982). For a partial list of the proceedings of the Austro-German Seminar see *Structure and Gestalt*, ed. Barry Smith (Amsterdam, 1981), p. iv—vii.

[17] On praxeology see Gunnar Skirbekk (ed.) *Praxeology* (Oslo, 1983).

[18] Kjell S. Johannessen, "Language, Art and Aesthetic Practice", *Wittgenstein, Aesthetics and Transcendental Philosophy*, eds. Johannessen & Nordenstam (Vienna, 1981).

[19] See Gunnar Danbolt's contribution to *Den estetiske praksis* (*Aesthetic Practices*) (Oslo, 1979), pp. 64—96.

CHAPTER VIII

DISCUSSING TECHNOLOGY – BREAKING THE GROUND

Strong claims have been made on behalf of the digital computer's capacity to surpass human beings as thinkers. It has been claimed that computers can eventually replace human thinking completely. The overwhelming philosophical difficulties involved in this claim have been acutely analyzed by Hubert Dreyfus in his book, *What Computers Can't Do*.[1] More recently, Joseph Weizenbaum, a distinguished authority on artificial intelligence, has carried the critique of computer simulation of human intelligence beyond Dreyfus. In his book, *Computer Power and Human Reason*,[2] Weizenbaum argues passionately and persuasively that there are potential catastrophes in the attitudes that the artificial intelligence community are fostering towards the digital computer in society at large. While I do not find all of their arguments equally convincing or well-formulated, I do not think that I am able to add significantly to them. This debate has raised nearly all of the significant questions facing intellectuals today. I would like to focus on some of the less discussed but none-the-less important issues before us: when should we abandon beliefs? and what are our obligations to convince our adversaries? I shall illustrate what I take to be a dangerous pitfall in discussions about technology by examining the history of a certain type of anti-technological attitude which existed in the Austrian periodical, *Der Brenner*.[3] To begin with, however, I must explain why I find the critique of artificial intelligence by Dreyfus and Weizenbaum convincing. Four considerations lead me to concur with them.

First, the discrepancy between promise and performance on the part of artificial intelligence seems to be prima facie evidence of what Imre Lakatos termed "sophisticated methodological falsification".[4] The most striking case in point concerns the assertions of Herbert Simon in 1957 that a digital computer would be the world's chess champion within ten years, that in the same time period a digital computer would discover and prove an important mathematical theorem and that most psychological theories would have the form of computer programs by 1967. The abject failure of these predictions in the time span allotted seems, at the very least, to call the feasibility of the project into question.[5]

The second consideration leading me to concur with Dreyfus and Weizenbaum is the notion of "essentially contested concepts" and the role they play in human life.[6] Basically, the most vexed issues which humans face involve conflicts about how we are to describe the situation we confront. For example, are we to admit that a urinal submitted as an art exhibit is art or not? Similarly, is abortion a routine gynaecological procedure or is it the denial of a fetus's right to life? These questions illustrate how our values are built into our descriptions of the world as well as the way in which those descriptions can be completely at loggerheads with one another. They also illustrate that it is entirely contingent whether we are able to resolve these conflicts or not. Conflict is the most real element in human life; to be human is to learn how to live with it. I do not think this is possible for a computer. In such situations it must either cease to function due to a lack of an effective procedure or reject one of the descriptions and, consequently, distort the situation.

The existence of *radical* conceptual change is the third consideration which leads me to consider the claims about the promise of digital computers to simulate human thought to be inflated. One of the ways in which the sort of conflict I have just described gets resolved (when it does manage to become resolved) is through a rethinking of the concepts which apply to some area of experience. So, when we decide that the urinal is art, we change the meaning of a term and reclassify our experience. Such transformations cannot be predicted or the result of some sort of cancatenation of pre-existing data because it is precisely the interpretation of the data that we are modifying. Radical conceptual change cannot be anticipated any more than a pun can be anticipated for to anticipate the pun is to produce it. The enormous complexity and the painful character of such changes have been recorded in such works as Thomas Kuhn's *The Copernican Revolution*,[7] Phillipe Ariés's *Centuries of Childhood*,[8] George Thomson's *Aeschylus and Athens*[9] and Loren Eisley's *Darwin's Century*.[10] In each case we confront a revolutionary (metaphorical) reordering of the universe of a sort which computers cannot perform.

Fourthly, Weizenbaum is correct to caution us that danger may lurk in the equation of human reason with computer cancatenation presupposed by the artificial intelligence research program. It is indeed *possible* that humans could only be able to understand what digital computers can understand, if we take the computer to be the model for all intelligence. I must add hastily that this danger does not lurk in the machine but in our expectation. I take Weizenbaum's point here to parallel that of critics of psychoanalysis who insist that being healthy mentally is by no means merely a matter of being "well-adjusted". There seems to be a rather arbitrary introduction of a standard of normalcy. Since there is no lack of data from contemporary sociology to indicate that our attitudes and assumptions about normalcy and deviance can create such behaviour patterns in the manner of a self-fulfilling prophecy,[11] we could indeed become dominated by our own instrumentalities.

These lines of argument lead me to reject the claims that digital computers can be produced which are capable of replacing human beings as reasoners. I must hastily add that what I have said in no way implies that computers are per se morally corrupting or socially useless. Computers are certainly much more efficient at cancatenating or performing effective procedures than humans are and certainly should be employed to do so. The complexities involved in implementing any notion of distributive justice on a worldwide scale may be greatly facilitated through computer power. I am asserting, then, that we should reasonably expect the claims which have been made to date on behalf of the ability of digital computers to surpass human beings in *all* respects to be modified in the light of performance to date and the inherent limits of cancatenating reason. This does not seem to be forthcoming. It prompts the question: when are we obliged to abandon a claim?

In the wake of Thomas Kuhn's *The Structure of Scientific Revolutions* philosophers of science have mulled over this perplexing question at great length.[12] Inasmuch as they have reached a conclusion, they have decided that there is no point at which it is possible to say that one must reject a paradigm or abandon a research program. The criteria for demarcating rationally justifiable from rationally unjustifiable programs are frankly prescriptive. Some methodologists have concluded from this that there ought not to be any efforts to distinguish what is scientific from what is not. Results alone would then determine what is and what is not a rational or legitimate research program. This would be one way of dealing with the issue, if it were not the case that a good deal of reasearch is carried on with public funding (though I tend to think that we would feel mighty uncomfortable to reject any *possibility* of being able to demarcate proper from improper usages of the term "science", or any other for that matter). It is hardly necessary to dwell upon the social consequences of funding anyone who came along with a research project simply because he happened to call himself a scientist. The results are too ludicrous to imagine. However, the question of public funding of research in an open (i.e. non-authoritarian) society does have far-reaching implications for the sort of situation I see artificial intelligence as confronting at the moment. In the face of powerful criticisms such as those which Weizenbaum and Dreyfus have levelled against artificial intelligence, it would seem that the public has a right to ask for an accounting which would justify its expenditure. This sum might, after all, be better spent, say, supporting the researches of unemployed philosophers. Thus, I think that Weizenbaum is justified in speaking of the issue before us as one of "civil courage". My remarks are certainly intended in this vein.

The problem before us, then, has to do with what it is for a question (or research program) to be legitimate. Such problems are inevitably burning ones for, as often as not, they are taken to imply that what is intellectually illegitimate is also in some sense or other immoral as well. I take it that there is a visceral sense of justification which arises simply with the continued existence of an institution or project over any considerable period of time. In this situation it is remarkably easy to consider critics as acting in bad faith. This may or may not be the case, but the assumption surely excludes the possibility of fruitful intellectual exchange (as we have seen in the United States in the debate over E. O. Wilson's *Sociobiology*). The strategy of ignoring one's opponent or of polemicizing against him actually serves to prolong rather than resolve the conflict. To resolve the conflict is to enter into a discussion with a view to convincing the adversary. This requires three things: both parties are acting in good faith, the assumption that it is at least possible that we may be wrong ourselves (fallibilism in the jargon of philosophers of science) and a willingness to meet our adversary on his own ground. For critics of technology this last requirement implies a willingness to address substantive technological developments. For the defenders of technology it involves willingness to participate in discussions of a philosophical, ethical or even religious nature. In short, both defenders and critics of technology must be prepared to leave familiar ground for the unfamiliar, seemingly irrelevant, territory from which our adversary's arguments have emerged.

I would like to illustrate my point by recounting the case of an attack upon the evils of technology which failed to make at least two of these assumptions. The article I want to discuss is called "The Technological Era". It appeared in the Austrian periodical *Der Brenner* in 1946.[13] It was written by a long (probably deservedly) forgotten writer by the name of Josef Bernhardt. From the perspective of 1979 "The Technological Era" could be a model of how *not* to write about technology. I hope to illustrate how a position we would almost universally find wanting today could be intelligible in the context of the history of the periodical. The latter will serve to cast light upon the way in which questions about technology, like other conceptual questions, are embedded in constellations of beliefs and cultural traditions but also help to explain why I consider it so important to meet our intellectual adversary on his own terms.

The most prominent feature of Bernhardt's polemic essay is its *religious* perspective. Bernhardt is concerned that modern society has deified technology with the result that modern men have a perverted relationship to nature and to work. He takes this perversion to be immanent in the three-fold project to which instrumental reason (i.e. calculation, *Zweckrationalität*, utilitarianism) has set itself: to *understand* natural processes for the sake of *predicting* its course with a view to *controlling* it. Thus, science is the presupposition of technology, which is the application of scientific knowledge and its *raison d'être*. The chief expression of the technological character of our age, Bernhardt asserts, is modern industrial production as exemplified by Taylorism and Henry Ford's assembly line. Like many other conservative social critics of his generation, Bernhardt laments the incursions of technology into areas that were the exclusive province of craftsmen within living memory. He considers that technology has polarized society into technocrats and machine wreckers, i.e. Communists and humanists respectively. This dilemma is lamentable but inevitable given the dynamics of the manipulative attitude to nature. Further, modern technology violates the Biblical injunction to regard nature as possessing moral value. These developments coincide with the emergence of modern society in the dissolution of the more 'natural' medieval order of things. The first signs of the emergence of the new era are perceptible in Hobbsean nominalism, late medieval conceptions of God as absolutely transcendent and ideas about popular sovereignty. Our problem today is that of mastering the mastery of nature. Finally, not the least of the charges which Bernhardt levels against technology is his accusation that it is responsible for the degeneration of language. This is tied to the phenomenon of the press and modern propaganda.

Three aspects of Bernhardt's essay are noteworthy for our purposes. First, he makes only a scant effort to convince his reader. The article cannot be said to contain arguments in any strict sense of the term; rather, it is a mere catalogue of the faults of technology. Secondly, for those of us who concur with Dreyfus and Weizenbaum, at least some of Bernhardt's opinions are justified, despite the fact that the main thrust of his argument is highly questionable. Third, given the traditions within which the *Brenner* operated, Bernhardt's opinions had much to commend them to its readers. To grasp the full import of the first and the last of these points we must glance briefly at the history of the periodical.

Der Brenner was founded in 1910 as a dissenting voice in the conflict between two varieties of rationalism, the theological rationalism of the Catholic Church and the scientism of liberal Pan-Germanism. It will be important to bear in mind that the journal always stood for tolerance and dissent throughout its forty-five-year history, even after it began its pronounced movement to the political right. Consciously emulating Karl Kraus's *Die Fackel*, its editor Ludwig von Ficker, and its chief contributor, Carl Dallago, aimed at defending the integrity of the individual aginst the onslaught of dehumanizing ideological abstractions. In today's jargon Ficker and Dallago aimed at nothing less than establishing an alternative paradigm for ethical behavior. They sought to substitute sensitivity for calculation. They sought to supplant the moral rhetoric of mass movements such as political Catholicism and Pan-Germanism with a concrete example[14] of pre-ideological, naturally moral humanity (the non-conformist Dallago supplied the requisite living example). Contemporary ideologies based upon the notion of 'progress through science' were among the forces which threatened the *Brenner's* Rousseauesque vision.

Specifically, Herbert Spencer's version of utilitarianism and Ernst Haeckel's Monism represented two versions of rationalism which explicitly accepted that the individual should be manipulated in the name of something 'higher'. Spencer advocated eugenic engineering for the lower classes. In Dallago's eyes this was little more than an assault on the very core of our humanity, our sexuality. The claims of Haeckel and the Monists were no less repugnant to him. Monism, too, aimed at nothing less than creating a 'scientific society', ruled, naturally enough, by a scientific elite.[15] The extravagant claims of the Monists did much to fix the attitude to instrumental reasoning in the German-speaking world. This is due, at least in part, to the fact that a number of eminent scientists were included among their number. Haeckel himself was one of the nineteenth century's foremost botanists. If anything, Haeckel was more responsible for the image of evolution and its cultural impact upon the nineteenth century than Darwin himself was. The Nobel prize winner, Wilhelm Ostwald, was another eminent Monist. The group took its name from its goal, the discovery of the essence of matter, the single element from which the whole universe developed. They claimed that this discovery would unlock *all* of the perplexities which had ever troubled men, even to the point where they would be able to realize the age-old Alchemist's dream of transmuting base metals into gold. The indispensable condition for doing so, however, was nothing less than the abolition of religion (except the religion of science, which they ordained as the public worship of the future). They were virulent chauvinists, it must be added, and the organization ultimately merged into National Socialism some twenty years later.[16]

Thus, the attitude which one finds to science and instrumental reasoning generally in the *Brenner* is less determined by science itself than the ideology of science. Many of the muddles about rationality in the journal are rooted in the fact that Ficker and Dallago did not distinguish the two but took the Monists at their word, i.e., as the single voice of rationality. Henceforth, the *Brenner* could not criticize rationalism on the basis of reason but had to seek an alternative that was 'irrational'. The only candidates which were available were art and religion. Art, however, was essentially narcissistic and, therefore, rootless and irresponsible. Indeed, the very foundation of the *Brenner* was as much a critique of aestheticism as it was a critique of ideologies and mass movements. In religion Ficker and Dallago found the *via media*, which was always to be the *raison d'être* of the periodical (though the understanding of religion would change drastically over nearly half a century). Nietzsche's Dionysian religion was precisely the sort of thing Ficker and Dallago sought; it emphasized spontaneity at the expense of reason. One could oppose both organized religion (which was no more highly esteemed in the early *Brenner* than it was by the Monists) and have something resembling a criterion of artistic and intellectual responsibility at the same time on the basis of Nietzsche's concept of religion. Thus,'the religious' was enlisted in the struggle against rationalism in the *Brenner*.

The periodical took a decidedly Christian turn when it was discovered that Christianity, too, has latent resources for social criticism which were at once radically individualist and implacably anti-rationalistic. This was a result of the rediscovery of the writings of Søren Kierkegaard and their translation into German. The *Brenner* played a significant role in the Kierkegaard revival by publishing many of Theodor Haecker's translations from the Dane's works.[17] These included Kierkegaard's assault upon the press and public opinion. "The Present Age". The dramatic advances of technology in this century tend to make us forget that developments in printing were among the most far-reaching technological innovations of the age. Further, the ability of the American press to expose the criminal deeds of the highest elected official in the land tends to blunt Americans in particular to the power of the press to distort what it presents the public. Thus, we can be shocked to discover that the activity of the press was also a source of anti-technological thinking in central Europe at the turn of the century. The exploitation of sexual scandals in the Viennese press was the epitome of the duplicity of the press to Kraus and the Brenner Circle. It was as if the press had to create them on its front page in order to deplore them on its editorial page. Further, the role of the press in providing propaganda during World War One was merely the culmination of surreptitious distortion of news on behalf of government and the industrial-financial complex. In short, the press appeared to be the instrument of manipulation *par excellence* to its critics, as much an infernal machine as thôse devised for waging war. Because the press was corrupt (as subsequent scholarship has verified), the very machine seemed to be implicated in its guilt. This leads us to another important factor in the development of anti-technological thinking in the *Brenner*, the notion that all issues are internally interconnected.

In the period in question it is extremely rare to find an outsider who actually was prepared to take each issue on its own merits. The conflicts of the area were preeminently conflicts of *Weltanschauung*. It is hardly possible to describe, let alone analyse, this phenomenon here. That would require nothing less than an account of the modernization of central Europe. Suffice it to say that it not only existed but became almost total in the Austrian state which emerged from the Treaty of

Saint Germain.[18] By 1920 Austria was a polarized society. The question of technology was even further ideologized. For the political right it was anathema (at least in theory), while the left embraced everything 'modern'. Thus, we find Bernhardt lamenting the division of the world into humanist and Communist. The *Brenner* made a valiant effort to transcend the socialist-clerical power struggle and even to mediate between the factions but this was impossible. Thus, by the time Bernhardt came to write his essay the issue of the nature of technology had very litlle to do with what machines actually do.

The importance of this story lies in the fact that it exemplifies what is sometimes referred to as the "embeddedness" of conceptual questions. They are always raised within context, from which they can be extricated only with the utmost difficulty. When there are reasons which are good reasons for accepting part of a constellation of beliefs (paradigm, research program), it is remarkably difficult to detach single elements from that constellation. Human beings are sufficiently clever to be able to find reasons for rejecting evidence which runs contrary to a deeply held belief especially when that evidence is presented from the standpoint of an alternative or rival constellation of beliefs. This says nothing about the truth or rationality of such a belief but merely about the ways in which beliefs operate in our lives. In the case of *Der Brenner*, some valid reasons for objecting to the inflated claims made on behalf of instrumental reason led to the establishment of a program which aimed at turning the tables on instrumental reason. Very early in its history, if not throughout it, opposition to the works of instrumental reason was detached from examining specific claims, theoretical commitments etc. Actually, opposition to calculation became an intellectual (spiritual) *practice* within a broader ethico-religious program. This helps to explain why we do not find anything like a rigorously developed argument in Bernhardt's essay. He was writing for an audience which had no doubts about the threats which technology represented to society. Two specific consequences of the attitude to instrumental reasoning which we find in the *Brenner* remain relevant to our discussions of technology today. Each illustrates how this attitude almost systematically undermined the *Brenner's* explicit intellectual and ethical goals.

In the first place, the program of polemicizing against calculation was itself at least partly responsible for the inability of the members of the Brenner Circle to recognize a profound critique of the intrinsic limitations of calculation. In the latter part of 1919 a manuscript was submitted to the Brenner Press, which claimed to set limits to the kinds of things which could be discussed meaningfully in any representational system. We know the book today as Wittgenstein's *Tractatus Logico-Philosohphicus.*[19] Wittgenstein submitted the book to Ludwig Ficker because he believed that he had shown once and for all that the logical foundation of language is such that it excludes any possibility of saying anything significant about ethical life. If he was correct, *Der Brenner's* program of opposing moral rhetoric with moral behavior was realized. The ethical quality of our actions is thus right there on the surface; it has nothing to do with the theories we appeal to, as the contributors to *Der Brenner* always insisted.

Wittgenstein claimed to have accomplished this feat by demonstrating that meaningful statements divide into tautologies and statements of observable fact. The former are substantively vacuous complexes, which make up the syntax of representational systems (languages). The only other possible ordered combination of symbols is a contradiction or logically meaningless statement. Wittgenstein devised the truth table method for representing tautologies, contradictions and statements of fact along a continuum. This facilitated understanding precisely how tautologies and contradictions were pseudo-statements. They were, as such, limiting cases along the spectrum of possibly ordered combinations of symbols. Though Wittgenstein regarded this as an achievement, he did not consider it to be as earth shaking as Bertrand Russell and the Vienna Circle did. By banishing unanswerable questions he did not see himself as ushering in the millenium, as the positivists thought, but, on the contrary, as showing "how little is accomplished when the problems are solved".[20]

Wittgenstein's view of his accomplishment in the *Tractatus* has implications for this symposium in as much as his work helped to make such things as digital computers possible. The ontology of the *Tractatus* is seriously defective as an account of the human world, as Wittgenstein himself was the first to see in a profound sense; yet, it remains a perfectly adequate delineation of a digital computer's "world". I do not wish to dwell on these topics, I am more concerned here with the fact that Ficker could not understand the *Tractatus* at all. It was a mystery to him that Wittgenstein could see himself as fulfilling the task that *Der Brenner* had set for itself. I take Ficker's rejection of the *Tractatus* to illustrate the danger of mere polemic rejection of instrumental reasoning. The members of the Brenner Circle knew so little about the substantial issues involved that an immanent critique of calculation was totally incomprehensible to them. Furthermore, they did not want to know anything about the subject. In some respects they seem to have shared the positivists' assumption that to understand logic, science, technology etc. is to be convinced of their virtually limitless powers. To avoid this the Brenner Circle found themselves on the verge of asserting the absurd notion that calculation could be eliminated altogether from human life. This muddle arose from taking the claims of the advocates of utilitarianism at face value. It never occurred to them that a Spencer or a Haeckel could be criticized from within, i.e. on the basis of reason itself. Consequently, they were incapable of *recognizing* a profound critique of reason, let alone producing *convincing* arguments, which might change someone's mind about the issues in question. They were in a conflict which they could never win. This situation led to the second important consequence of *Der Brenner's* attitude to science, frustration of its own commitment to communication.

I have emphasized Ficker's desire to find a *via media* in the midst of ideological conflicts from the very foundation of his periodical. This dedication grew from 1919 when Ficker made the acquaintance of Ferdinand Ebner. The latter became a chief contributor to the periodical till his death in 1931. In his *magnum opus, Das Wort und die Geistige Realitäten* (the work that Ficker chose to publish in rejecting Wittgenstein's *Tractatus*), Ebner emphasized that we are only human to the extent that we recognize the humanity of others.[21] Recognition takes place when we speak to them. Thus there is no I without a Thou but also no personality without communication. Thus, *Der Brenner* came to be committed to a philosophy of dialogue. Its tragedy lies in the failure to recognize that dialogue is meaningless without criticism. Criticism, however, must be mutual. Ebner thought that his discovery of the dependence of the I upon the Thou (he wrote before Martin Buber) entailed yet another critique of theorizing reason. It is certain that dialogue will hardly be possible when the parties involved do not speak the same language, as Ebner insists. However, this would seem to place an obligation to learn the language of the other for the sake of dialogue, which exists on the part of the individual who is committed to dialogue. The latter certainly cannot expect that from the former. However, the attitude that *Der Brenner* assumed towards calculation precluded exactly the sort of dialogue that its commitments required.

At this point we can draw the moral from the story. The history of the controversy over technology shows that there has been closed-mindedness on the part of the opponents of technology. Without making any accusations, I simply want to suggest that there just *might* exist a corresponding closed-mindedness on the part of its advocates. If there can be an anti-technological intellectual practice, it is at least possible that there can be a pro-technological intellectual practice that systematically ignores its adversary. The danger in this attitude is that we can become encapsulated within our own thoughts, on the one hand, and that we virtually give our opponents *carte blanche* to ignore us, on the other. This is to abandon the hope of winning the conflict. *Der Brenner* episode illustrates that subjective certainty, the ability to appeal to evidence and tradition do not guarantee that our opinions are justified. Our evidence may be the wrong sort of evidence and our tradition may lead us to ask the wrong questions. We must be at one and the same time guided by what we take to be the substance of the issues at hand and also prepared to reconsider precisely what the substance of the issue actually is. It is always necessary to bear in mind that the most serious issues we confront concern "essentially contested concepts", i.e. disagreements over just what the substantive issues are. To prevail in the conflict is to be prepared to follow the discussion, even when it leads us into unfamiliar terrain. To fail to do this is to sell our own intellectual positions short but also, to fail to live up to the demands of courageous citizenship.

NOTES

1 Hubert Dreyfus, *What Computers Can't Do* (New York, 1972).

2 Joseph Weizenbaum, *Computer Power and Human Reason* (San Franciso, 1976).

3 On Der Brenner see Walter Methlagl, "*Der Brenner 1910—1954: Leben und Fortleben einer Zeitschrift*", *Nachrichten* aus dem Kösel Verlag (München, 1965) and Gerald Stieg, *Der Brenner und die Fackel: Ein Beitrag zur Wirkungsgeschichte von Karl Kraus* ("Brenner Studien" Bd. III; Salzburg, 1976).

4 On the concept of "sophisticated methodological falsification" see Imre Lakatos, "Falsification and the Methodology of Reasearch Programs", *Critcism and the Growth of Knowledge*, ed. Imre Lakatos and Alan Musgrave (Cambridge, 1970), pp. 91—196.

5 Dreyfus, p. xxix—xxx.

6 The notion of an "essentially contested concept" was originally developed in his article "Essentially Contested Concepts", *The Importance of Language*, ed. Max Black (Englewood Cliffs, NJ, n.d.) pp. 121—46. For a penetrating analysis of the role such

concepts play in human affairs see William Connolly. *The Terms of Political Discourse* (Lexington, MA, Toronto and London, 1974).

[7] Thomas S. Kuhn, *The Copernican Revolution* (Cambridge, MA, 1957).

[8] Phillipe Aries, *Centuries of Childhood*, trans. Robert Baldick (New York, 1962).

[9] George Thomson, *Aeschylus and Athens* (New York, 1968).

[10] Loren Eisley, *Darwin's Century* (Garden City, NY, 1961).

[11] The essays in Stanley Cohen's anthology *Images of Deviance* (Harmondsworth, 1971) bear this out. Jock Young's "The Role of the Police as Amplifiers of Deviance, Negotiators of Reality and Translators of Fantasy" pp. 27—61 is an especially noteworthy illustration of my point.

[12] The essays in *Criticism and the Growth of Knowledge* present several of the most widely accepted views of the issue. See also Stephen Toulmin, *Human Understanding*, Vol. I (Princeton, NJ, 1973).

[13] Josef Bernhardt, "Das technische Zeitalter", *Der Brenner* XVI (1946) pp. 73—101.

[14] On Carl Dallago see my article "Carl Dallago and the Early Brenner", *Modern Austrian Literature* **11** (1978), pp. 1—17.

[15] On monism see Daniel Gasman, *The Scientific Origins of National Socialism* (London, 1971).

[16] After completing this essay I found the article by N. R. Holt, "Monists and Nazis: A Question of scientific Responsibility", *Hastings Center Report* 5 (1975), pp. 37—43 which seemed to contradict the account of the matter in Gasman. Holt asserts, finally in 1933, when many German organizations were ideologically "co-ordinated" with the new ruling National Socialism, the membership of this major popular scientific organization chose to disband. However, Holt neglects to state that a singificant element within the membership of the defunct Monist League then banded together under the aegis of the Nazi Gauleiter of Thuringia to form the Ernst Haeckel Gesellschaft. See Gasman, p. 173.

[17] On Theodor Haecker see Eugen Blessing, *Theodor Haecker* (Nürnberg, 1959).

[18] Some of the aspects and implications of the polarization of society in the First Austrian Republic are discussed in Alfred Diamant, *Austrian Catholics and the First Republic* (Princeton, NJ, 1960) and in Klemens von Klemperer's penetrating study, *Ignaz Seipel: Christian Statesman in a Time of Crisis* (Princeton, NJ, 1972).

[19] See my article "Wittgenstein, Ficker and the Brenner", *Wittgenstein: Sources and Perspectives*, ed. C. Grant Luckhardt (Ithaca, NY, 1978), pp. 161—89.

[20] "Wie wenig damit getan ist, dass die Probleme gelöst sind", Ludwig Wittgenstein, *Tractatus Logico-Philosophicus* (Frankfurt/Main, 1976), Vorwart.

[21] On Ebner see Theodor Steinbüchel, *Der Umbruch des Denkens* (Regensburg, 1936) and Stieg, pp. 203—34.

CHAPTER IX

SOCIALIZATION IS CREATIVE BECAUSE CREATIVITY IS SOCIAL

Nearly everything about the approach to creativity and socialization based upon positing a strict dichotomy between them is misconceived. The mode of questioning, deployment of examples and comparative procedures systematically insure that such an approach will distort precisely what it seeks to clarify. Thus it is not that there are no genuine problems, tensions and paradoxes surrounding creativity and socialization in the social research context; but that the dichotomizing approach is far too abstract to illuminate those problems, tensions and paradoxes. The reasons why this should be so are, nevertheless, far from uninteresting. This is because they are bound up with central issues relating to the ways in which language and action are interwoven.

The thesis that I want to call into question lies at the heart of the matter. It is the claim that socialization is inherently conservative, i.e. in all respects stifling of creativity.[1] Posting a rigid dichotomy between socialization and creativity is little more than considering socialization as mere regimentation at the same time that we stipulate that creativity is nothing less than *creatio ex nihilo*. Thus from the dichotomy between stifling regimentation and total innovation there emerges a dilemma according to which creativity is taken to be inextricably bound to uniqueness and ambiguity; whereas socialization is no less inextricably bound to transmitting conventional wisdom in a manner which necessarily stifles originality.

I contend, to the contrary, that this alternative has been arbitrarily concocted in such a way that it leaves us no elbow room. It traps us between a crude insistence, at once nominalistic and Romantic, upon the unfathomability of individuality and an equally crude, rationalistic conception of communication as an iron cage. I suggest that these descriptions of the moments in what ought to be termed learning do not emerge from its analysis but are first projected upon it and then reinforced by bad examples (when there are examples) and invidious comparisons, such that what begins as conceptual tragedy culminates in rhetorical farce as the boundaries between the actual conditions (i.e., historical and sociological) under which we might happen to learn in some situations are hopelessly confused with the transcendental conditions for learning itself.

The best place to begin to untie the conceptual knots with which this dichotomy binds us is by posing a series of questions about the assumptions which are built into that dichotomy: "what is it that is so good about creativity or innovation?", "what is it that is so bad about socialization within a disciplinary matrix?". For the dilemma with which we have been confronted to make sense in the first place the answer to these questions must be unequivocal. Innovation has to be a value in and of itself: what is new is good; what is old is, if not bad, at least suspect. What stifles the imagination in the production of novelty, above all tradition, is mere routinization and as such odious. Thus it follows that positing a sharp dichotomy between creativity and socialization entails a commitment to "modernism" or avantgardism in the sense that what is new must necessarily be good. That sort of claim is only interesting to the extent that it is a strong claim about the absolute superiority of novelty. But to have made such a claim is by no means to have sustained it. It is only possible for the dichotomist to sustain this claim if he can in some way demonstrate that what is new is by that very fact an accomplishment in some non-trivial sense of the term. This I take to be all but impossible. Counterexamples to that thesis abound. For example, C. W. von Naegeli (to whom I shall return below) produced a theory of evolution based upon the notion that the material basis of heredity was viral and according to which the most complex animals are the oldest and the simplest, e.g. the amoeba, the newest. This conjecture was surely original but it was hardly an achievement. It should be clear that such examples from the history of science could easily be multiplied. However, to say this is to make the question of novelty — or 'creativity' as it is understood by the dichotomist — secondary to that of achievement. To be sure, scientific achievement entails innovation, but the relation is not reciprocal. To be creative implies more than being anti-traditional or avantgarde; it implies substantive accomplishment in terms of criteria for the growth of knowledge of precisely the sort that a discussion of learning in terms of the dichotomy creativity/socialization obscures.

Furthermore, it focusses our attention falsely in the sense that positing that dichotomy seems to involve positing the dubious thesis that creativity is one thing. It is no small part of Ludwig Wittgenstein's service to contemporary thought to make us skeptical with respect to the isomorphism between words and things. To discuss creativity profitably we need examples of just what it might mean in the research context to refer to a piece of work as "creative". The examples that we have been given by Olsson are mostly drawn from the areas of art (Duchamp) or literature (Mallarmé, Joyce). They certainly merit discussion but there is much to be said for the view that they are irrelevant to a discussion of social research. At least the burden of proof lies with him who would introduce them as example to explain how they illuminate the research context. It is most unclear what is would be like to be, say, the Joyce of American political science. This is in no small part due to the fact that, for all his creativity, Joyce's work has had a relatively small impact upon the practice of writing novels in the English language. I do not mean to disparage either Joyce's achievements, which are enormous, nor literary scholars' estimates of them, without which Joyce would be mighty difficult to fathom, but to suggest that it is unclear what Joyce has meant as an exemplary writer in a way that it is clear what, say, Henry James and William Faulkner have meant to that enterprize.[2] Joyce's very creativity makes him an all-but impossible model to follow but that does not imply that there are no models that subsequent authors have followed without in the least compromizing their own originality. In short, Joyce would seem to be a very speical case. To use Joyce normatively in a discussion of what it is to be creative is to be at the very least highly selective, if not idiosyncratic. The point is that there are creative individuals and creative individuals (Joyce *and* Henry James); but we learn more about how to create from some than others. Moreover, it is a wholly gratituitous assumption that they all share some specific characteristics (just as the obverse, i.e., that they have absolutely nothing in common would be equally gratuitious).[3] Indeed, the idea that there are interestingly common features to innovation in, say, chemistry and art history is as implausible as the notion that the incommensurability between them excludes the possibility of saying anything about the subject. What we need is examples of the genesis of recognized achievements Fixating upon the dichotomy between creativity and socialization would seem to focus our attention away from the direction in which we might expect to find illumination with respect to this issue. It is no less important to emphasize that the very fascination with novelty, which the creativity/socialization dichotomy highlights, has been taken by at least some philosophers to be the most dubious element in modern culture. Here again Wittgenstein comes immediately to mind. In his concern for establishing the conceptual limits to human endeavor, limits that the very nature of our language would tempt us to transcend, Wittgenstein saw himself as radically out of step with a society whose very *form* was progress. He contrasted his work with that of his more typically modern contemporaries by describing the latter as an obsession with erecting new structures; whereas at the same time neglecting the foundations of existing ones.[4] For Wittgenstein this was little less than a symptom of the modern spiritual malaise. Its main feature is the unrestricted desire to possess or to abolish all limits. One of its manifestations — albeit one which Wittgenstein did not discuss — is the desire to be original. In the world in which Wittgenstein grew to maturity even more than in our own nothing was more insulting than to be termed a mere epigoni. Nobody has better expressed the paradoxical result of this fascination with the new, with its emphasis upon creative individuality, upon letters than E. M. Cioran. Cioran once observed that in the eighteenth century writers strove for a common clarity and elegance of expression, rather than an individual style, with the result that the eighteenth century saw a flourishing of individual style that has rarely been equalled; whereas our century, for all its worship of the uniqueness of the individual, has been one in which writers increasingly resemble each other stylistically.[5]

The tendency of working artists to avoid discussions of creativity should not pass unnoticed here, for it tells us something about the utility of such a discussion. For example, painters, following the lead of Picasso, who insisted, "il est défendu à parler au pilot", especially tend to avoid such discussions.[6] The successful artist accomplishes something concrete with color, line, texture or imagery, which can only be discussed against the background of previous efforts to do something analogous. I take this reticence on the part of painters to be rooted at least in part in the tediousness of explaining the background to their work to the uninitiated. In short, either one knows about the development of the movement to which a painter belongs as well as her individual development or one does not. Nothing could be more boring — or indeed more difficult — for the artist than to explain what she takes for granted. It is no less important to point out that in the eyes of working artists the achievements that the public most admires are principally seen as challenges, as points of departure, inasmuch as they suggest new problems.[7] Any discussion of the structure of an artist's problem is bound to strike us here as weary, stale, flat and unprofitable. How much more, then, will the discussion of an abstraction called "creativity" be superfluous in this context. I submit that the same is true of achievements in research: what is interesting and important about it will have more to do with the substance of achievements involved, and thus be inextricably bound to a specific *Problemstellung*, than it will with something called "creativity". Indeed, I am inclined to think that in both painting and in research discussions of creativity put the matter of assessing the significance of a given accomplishment at a hopelessly general level. One important implication of all of this is that an understanding of these practices requires what could be called hermeneutic reconstruction, i.e., an informed account of how specific accomplishments come into being.

The obverse of the point that discussions of creativity put matters at a hopelessly general level can be made by looking at the example of what might be called the creative speculations in the history of genetics between Darwin and Mendel, when all sorts of highly innovative views about the material basis of heredity were advanced by scientists like G. W. von Naegeli, often without the slightest empirical foundation.[8] To focus upon the creative aspects of those speculations is to make a fascinating investigation into the development of the biological imagination, but one which tells us precious little about the scientific achievements of those speculators. To do that we have to tell the whole story of the development of the discipline from Darwin to Mendel. To think otherwise would be to assert that we can understand examples without knowing what they are examples of. This would be to put ourselves into the position of the woman in Chuang Tsu's parable who wanted to learn the secret of being beautiful.[9] To this end she visited a woman fabled for her beauty. However, upon her arrival the legendary beauty was for whatever reason wholly distrought and crying. From this the naïve visitor concluded that all she needed to do to be beautiful was to cry and beat her breast. Clearly she missed the point. Positing a strict dichotomy between creativity and socialization invites us to do just the same.

What about the other half of our dichotomy? What about socialization? Just how anti-innovative is it? To be sure to learn, say, to play chess, to paint a picture or to study electoral behavior involves mastering a set of techniques. That this is often tedious is beyond question; that in the case of research training the whole process can be mind-destroying is equally plausible, the question is must it be so? When we pose the question of the relationship between creativity and socialization and creativity in terms of a strict "either/or", I fear that it must be; but I must hastily add, because it posits as well (1) a number of abstract assumptions about the nature of language, discourse and learning, which forces this conclusion on us and (2) a disdainful attitude to concrete examples of outstanding research and, ultimately, to the logic of research.

The pernicious assumptions about language built into the creativity/socialization dichotomy are: (1) the notion, common to Carnap and Saussure, that language is a code, (2) its corollary that knowledge is translation, (3) the notion that communication is a prison (presumably because it locks us into the logical structure of a code) and that knowledge must be cumulative. All of this depends upon the conception of language as a logical *system*, which is to say that the actual structure of language is not displayed in its use (*parole*), but only in its formal or syntactical representation (*langue*).[10] But it is precisely this picture of language as a cage that pragmatically inclined philosophers such as Wittgenstein have called into question. It would hardly be possible here even to sketch all of the nuances of that pragmatic alternative to the positivist/structuralist picture but it is at least possible to trace the main features of that perspective on language.

To the allegation that language is a code and that knowledge is a matter of translation and the like we can respond with the Lubasz Gambit (after Harry Lubasz my old professor of social theory, who perfected it). This is to respond, once confronted with an implausible, ludicrous or absurd thesis: how did you find out?[11] The point of introducing that move here is to underscore the fact that we only know what a code is on the basis of ordinary, pre-theoretical language. A code is a way of employing language, a part of language, which we can recognize because we can distinguish codes, encoding and the like from other activities and parts of language. To say that language is a code — and that knowledge is a matter of translation etc. — is to employ a curious metaphor, which likens the whole to a part and then suggests that we should forget the other part. It is roughly like saying that football is kicking and going on to analyse football on the basis of what it is to kick alone, i.e., excluding strategy, defense etc. It can be done but it is radically insufficient as an explanation of the game. It is curious to insist that language as such is a code because this is certainly not something that we are conscious of. But, given that this is the case, how could we ever discover that language is a code? How could we get outside of language to do so? The simple answer is that we cannot. No, it is only meaningful to call something a code to the extent that we can distinguish it from those things that are not codes. Thus, to say that language is such as a code is to say something highly confused. Most of the trouble with the approach to creativity and socialization to which I object has its roots in this confusion.

Once we have acquiesced to the view that language is a code, or a formal syntactic system, it is a fairly short spring to the notion that dialogue is a mode of imprisonment. Dialogue becomes something "greedy and cautious",[12] not because that is the way people in dialogue actually relate to one another, but by virtue of the way in which speaking a language involves commitment to a syntactic system. On this account, it is easy to see how monologue can be conceived as superior to dialogue inasmuch as it does not involve such a commitment and can thus be characterized as "generous and open". Indeed, on the basis of the views of philosophers as different as Wittgenstein and Bachelard there is much to be said for the idea that the form of philosophical reflection is monologic. However, there are serious prima facie reasons for rejecting the sort of contrast between dialogue and monologue which follows from the positivist/structuralist picture of language. These objections stem at once from psychoanalysis and the personalistic philosophy of Ferdinand Ebner.[13]

Both identify mental illness as a disturbed mental state, which Ebner has aptly described as *Ich-Einsamkeit*, isolation of the ego or encapsulation within the self. They remind us that, although monologue may be brilliant, it always, by its very nature, runs the risk of falling out of contact with reality. Monologue, after all, is the form of life of the psychotic. The importance of dialogue in human life follows from the danger that we loose the ability to distinguish fantasy from reality. Put differently, the importance of dialogue follows from its nature as *critical interaction* with the other. Dialogue is not simply a matter of talking, but of taking the other's conversation *seriously*. I know that I have done this only when I come to realize that I might be mistaken in my views as a result of taking the position of the other seriously.[14] I only know that I am in contact with another human being when I am aware that the other has convinced me to re-evaluate some aspect of my beliefs. For this reason, Ebner rightly insisted that spiritual reality could only be spoken of in the plural and as coming into being through speech. Thus the title of his book: *Das Wort und die geistige Realitäten*.

Here it is worth pointing out that Wittgenstein's philosophical monologues, unlike Bachelard's, take the form of a *soliloquy* or a dialogue of the soul with itself as St. Augustine described that genre, whose point is self-examination or self-criticism. If philosophical problems are the result of dream-like fixations according to which we nourish ourselves with a one-sided diet of examples, philosophical reflection dedicated to dissolving those fixations must take the form of self-interrogation through which we establish distance from ourself. In any case, two people who always agree are by no means on that account partners to a dialogue. Precisely because dialogue only has meaning as critical transaction, it is wholly inappropriate to describe it as "cautious and greedy". The crucial point is that the critical constraint that the partners place upon one another — and it is important to see that the partners in the dialogue are in a position to place critical constraint upon one another because they are each prepared to take a perspective other than their own seriously — *is* the very act through which objectivity is constituted. Thus, whereas dialogue may constrain, it is a mistake to see it as imprisoning.

This brings us to the very center of my criticism of the creativity/socialization dichotomy: it can only be taken seriously if we consider socialization in the context of research training as mere regimentation or indoctrination. Now, I have no doubt that a lot of what passes as research training — and many other socializing institutions as well — has this character, but I see no reason to think that it *must* always and everywhere be so. Indeed, the best research training would appear not to have that character. Moreover, the central notion in Wittgenstein's later philosophizing, the notion of following a rule, would seem to indicate that a 'creative' moment is at the very center of 'socialization'. It is these themes that we now must explore.

The place to start is with the logic of research. By this term I mean neither its formal logic nor Popper's logic of scientific discovery but, rather, the working logic within disciplines, their modes of advancing assertions, marshalling evidence, inferring and justifying their modes of inference. To be enculturated into a discipline is to learn not simply the current state of knowledge regarding a certain subject but also to learn how to speak and write about that subject. That is, after all, why we have seminars and term papers. It is by means of 'practice' of this sort that we master — and display that we have mastered — the working logic of our discipline. Lest this be confusing, I do not mean to assert that this working logic is ever explicitly formulated as such; for it rarely is — indeed, with John Dewey and my colleagues in the Bergen School of Aesthetics, I take it to be the task of the philosopher of science to perform just such a task as reconstructing that logic, which forms the basis of day to day work in a field of research.[15] This is the sort of thing one learns as one masters a field. For example, in the course of becoming an historian one learns that diplomatic histories which are exclusively based upon official documents are not to be trusted; for what is crucial for understanding diplomatic exchanges are the assessments of the various participants with respect to their own situation.[16] This requires biographical and autobiographical information, which gives us the basis for formulating their assumptions, and which alone provides the historian with the basis for assessing achievement or failure with respect to a diplomatic mission. Thus, a study in diplomatic history could be a painstakingly thorough analysis of all official communiqués relating to a given diplomatic exchange and, yet, be radically deficient because it has brought the wrong sort of information to bear upon the case. Here it is not that the data are wrong, but that the wrong sort of data have been appealed to. To become an historian is to learn such principles; they provide one with criteria of relevance and, therefore, criteria for evaluating the *methodological* soundness of work in the field.

To be sure, this sort of critical skill is compatible with a disciplinary blindness or narrowness which may contingently be present. The point is that status within the historical profession — or any other, for that matter — accrues inter alia to critical skill. However, that can only be learned through criticizing specific works.[17] One 'picks up the knack' only through practice. This is the meeting point for socialization and creativity: critical activity. In the end, one learns to ask why it is that a given belief is simply assumed within a discipline often with astonishing results; for occasionally, upon examination, it turns out that unquestioned assumptions within a discipline simply do not bear up to questioning. This has been recently driven home to students of twentieth century German history, for example, in the brilliant work of Richard Hamilton. Till 1982, when Hamilton published his superb study, *Who Voted For Hitler*?, students of German history believed unquestioningly that Hitler's chief electoral support came from the lower middle class.[18] Fifty years of research was informed by this assumption, which had, from the start, never been questioned, when it was in fact relatively easy to do so. On the face of it, this might seem to give aid and comfort to the defenders of a sharp division between creativity and socialization in research, i.e., the fact that it took fifty years to challenge this assumption. Moreover, the fact that Hamilton is a sociologist, and, therefore, not strictly speaking an historian might tend to reinforce this view. However, objections of this sort can be met. First of all, the German history in the twentieth century has been a subject which has been consistently investigated for decades by sociologically oriented historians and historically interested sociologists. Thus, the interdisciplinary perspective was already accepted within the area of research and was by no means unique. As to the length of time involved, it does indicate that there are problems about socialization within disciplinary matrices — problems, by the way, about which Hamilton has some very interesting things to say — but the point is that the time-honored assumption under discussion was called into question from *within* the disciplinary matrix on the basis of reasoning which was by no means radically new. The point is that a brilliant work of 'creative' scholarship can take its point of departure from a single question, when that question happens to involve an unquestioned assumption. That which determines the 'creativity' of Hamilton's work is in no sense an arcane skill — even if it certainly requires courage to ask such a question in the face of universal agreement, this because it will surely seem silly to dispute what everybody 'knows'. Socialization within a discipline or a disciplinary matrix, since the question of what constitutes a discipline is with a few exceptions, usually disputed these days, is not as such incompatible with creativity, even if it often is in fact. It is, then, a contingent matter to what extent a given discipline at a given time is open to radical innovation in thc form of taking seriously new classes of facts. It seems that German history can do this in a way that archeology, at least as it is currently practiced in the United States, cannot — as the experience of a sociologically-minded ancient historian of my acquaintance there would indicate. My main point here is that the politics and sociology of disciplinary matrices may tend not to favor conceptual innovation for all the reasons that Max Weber drew our attention to long ago; but that is in and of itself insufficient to say that organized structures for producing and transmitting knowledge have to be opposed to innovation as Weber's own substantive achievements would indicate. As long as criticism is part of socialization within disciplinary matrices, there is no reason to posit a strict dichotomy between creativity and socialization. Moreover, it seems reasonable to think that methodological self-reflection, i.e., philosophy of science construed as Deweyite reconstruction of actual practices within the discipline and interdisciplinary perspectives are two modes of furthering the requisite self-criticism for creative socialization within research programs. However, it should be clear that there are no guarantees that this will be the case.

Here is perhaps a good point to bring the *two* aspects of innovation fully into our discussion. Conceptual innovation requires both altering basic models, metaphors and examples, which we employ to articulate the structure of our problems, but it also requires that the community of scholars be prepared to accept new models, metaphors and examples. It is not enough to suggest, say, that electrical phenomena should be modelled on the analogy with the current of a stream; it is equally necessary that the scholarly community be prepared to take a look at the matter from that perspective. Now, it is precisely critical discussion or dialogue which distinguishes a Mandarin caste from a scientific community. It is only when a putative innovation is subject to critical scrutiny that the value of that alleged innovation can be determined. Thus, criticism becomes the prerequisite for growth of disciplines inasmuch as it binds context of discovery to context of justification. Because there is a difference between innovator and crank, between pure fancy and disciplined speculation, it is necessary to insist that genuine creativity is something fundamentally social.

Thus, we discover the dialectical identity of creativity and socialization. I want to conclude by indicating how the central insight in Wittgenstein's later philosophy concerning what it is to follow a rule is in its way also an articulation of this dialectical identity. Since the implications of this notion are not at all well understood, it will do well to indicate how *not* to understand Wittgenstein on rule following. Wittgenstein is sometimes understood as giving aid and comfort to a certain type of conservatism, which would seem to support a sharp dichotomy between creativity and socialization, when he insists that our understanding of the world is the result of training and that this training entails following rules *blindly*.[19] However, this is not the case. When Wittgenstein makes this assertion he intends to emphasize that, given the social nature of rules, we do not have a choice with respect to the 'internal order' of the action stipulated by the rule. That is determined in advance. Thus, in playing chess, for example, the Bishop must move diagonally only. If I move it any other way, I cease to be playing chess. However, there is nothing to say that I cannot invent a chess-like game in which Bishops also move, say, one square in any direction, i.e., like the King — nothing that is except my fellow chess players and the other rules of chess. I can only make this innovation, if someone will play that way with me, which, in turn, depends upon both the willingness of the other party *and* the compatability of my new rule with the established way of playing chess. It may well turn out that my putative rule will not 'nest' with the established rules, that the result will be hopelessly confusing or destroying of the logical grammar of the game. Furthermore, even sticking to the rules of chess as normally followed, my following the rule blindly is not incompatible, with the idea that there are many ways of following the rule. It is essential to rule following behavior for Wittgenstein that it be open-textured despite being socially conditioned and internally ordered.

This is not a mere stipulation on Wittgenstein's part, but follows from the central role of *guessing* in rule-following behavior and a-fortiori in all learning.[20] For Wittgenstein teaching is not a matter of passing on information in some kind of rote manner, but a training in the application of rules. To be able to apply a rule is simply to know what to do next. But to do this in playing chess, distinguishing subtle shades of color or doing mathematical puzzles is to make a guess on the basis of the examples we have been given and to continue to do so till we have found the 'knack' for doing so correctly. It is an eminently *analogical*; rather than *digital* activity. Thus, Wittgenstein insists that when I teach somebody something I teach him *everything* I know, but that he genuinely learns what I teach in the sense that he appropriates that knowledge wholly in his own way.[21] Socialization, or learning to follow a rule, then, is in itself anything but regimentation. Once more, this is not to say that it cannot become so, but that it is not determined to become so. If rule-following does indeed become a regimentation that is a consequence of the socio-political context; rather than of the 'ontology' of rule-following. This leaves us with the paradoxical notion that learning is social, i.e., because I must learn from others, but that it is also 'creative' to the extent that it is essentially a metaphorical, rather than a rote activity. So, if we must be saddled with a discussion of 'creativity' with respect to research or anything else for that matter — and I hope it is clear that I do not think that we *must* — we can conclude that creativity is social, because socialization is creative.

NOTES

[1] This paper is a critique of some central ideas of Gunnar Olsson, Director of the Nordic Institute of Urban and Regional Planning, in particular his "Of Socialization and Creativity", Nordiska institutt för samhällplanering Meddelande, 1984, no. 7. (Stockholm).

[2] I owe this point to Professor John Keenan of the English Department of LaSalle University, Philadelphia.

[3] See Stephen Jay Gould, *The Mismeasure of Man* (New York & London, 1981), pp. 252—255; pp. 296—316.

[4] Ludwig Wittgenstein, *Vermischte Bemerkungen*, ed. G. H.von Wright (Frankfurt/ Main, 1977), p. 20.

[5] E. M. Cioran, *The Temptation to Exist*, trans. Richard Howard (Chicago, 1968), p. 131.

[6] Cited in Tore Nordenstam, *Explanation and Understanding in the History of Art* (Bergen, Norway, 1978), p. 91.

[7] I have profited very much from discussions of artists perception of their own work and willingness to talk about it with Gregor Traversa and Melissa Zeppa.

[8] Ernst Mayr's discussion of Naegeli's 'contribution' to modern biology is a case in point, *The Growth of Biological Thought* (Cambridge and London, 1982), p. 671f.

[9] Chuang Tsu in Arthur Waley, *Three Ways of Thought in Ancient China* (Garden City, nd.), p. 20.

[10] Jonathan Culler, *Saussure* (London, 1976), p. 29ff.

[11] See Heinz R. Lubasz, Professor of Social Theory, University of Essex, Colchester, England.

[12] Olsson, *op. cit*.

[13] Ferdinand Ebner, *Das Wort und die geistige Realitäten* (Frankfurt/Main, 1980), p. 262; cf. Klaus Dethloff, "Ferdinand Ebner und die Psychoanalyse oder Traum vor und nach dem Einschlafen", *Gegen den Traum vom Geist*, eds. W. Methlagl et al. (Salzburg, 1985), pp. 162—73.

[14] See my "Popper und Ebner als Denker", *Gegen den Traum vom Geist*, pp. 25—32.

[15] See John Dewey, *Essays in Experimental Logic* (Chicago,1916). For the Bergen School's approach to the reconstruction of aesthetic practices see the contributions by Gunnar Danbolt, Kjell S Johannessen and Tore Nordenstam in *Contemporary Aesthetics in Scandinavia*, eds. L. Aagaard-Mogensen and G. Hermerén (Lund, 1980), pp. 81—132. For Stephen Toulmin's development of Dewey's 'experimental' logic see *The Uses of Argument* (Cambridge, 1957) and Toulmin, Rieke and Janik, *An Introduction to Reasoning* (2nd ed.; New York, 1984).

[16] I am indebted to Professor Kenneth Barkin of the University of California, Riverside for my understanding of conventional historical practices.

[17] Here again, I owe much to Kenneth Barkin; cf. my collection *How Not To Interpret A Culture* (Bergen, Norway, 1986).

[18] Richard Hamilton, *Who Voted for Hitler*? (Princeton, 1982).

[19] Wittgenstein, *Philosophische Untersuchungen*, **I**, p. 219.

[20] *Ibid*., **I**, §210.

[21] *Ibid*., **I**, §208.

CHAPTER X

MYTH AND CERTAINTY

The idea that Wittgenstein's *On Certainty* contains important insights for understanding Georges Sorel's concept of social myth will surely strike a lot of people as implausible, ludicrous or absurd. So, it is necessary to begin by explaining the point of this exercise. It is certainly intended neither to put old wine into new bottles nor to open up a new branch of the burgeoning Wittgenstein industry. It is, rather, an effort to do three things. First, reading Sorel through Wittgensteinian spectacles should help to rescue his insights, perhaps in a way that he would challenge, given the opportunity, from the confusions arising out of the woefully inadequate Bergsonian conceptual framework in which they were developed. The point of departure for this is taken from H. Stuart Hughes's classic study, *Consciousness and Society*, perhaps the most important presentation of Sorel in America to date. Hughes writes: "Sorel never found a proper vocabulary nor a suitable conceptual scheme into which he could fit what his critical intelligence had taught him".[1] Secondly, my inquiry will provide an occasion for further articulating some of the implications of Wittgenstein's later philosophy for social thought. Since social scientists in increasing numbers have found insight and stimulation in Wittgenstein's work — and, since Wittgenstein himself said virtually nothing on the subject — the procedure of using Wittgenstein to rescue Sorel is equally important as a part of those efforts to appropriate Wittgenstein's insights for those concerns of ours which they promise to illuminate. Thus, it is less a matter of "how to do things with Wittgenstein" than it is of philosophizing in Wittgenstein's spirit. A third reason for undertaking to 'rehabilitate' Sorel on the basis of Wittgenstein has to do with our getting straight about the nature and scope of social thought. To get "straight" about the nature and scope of social thought here means to obtain an illuminating perspective upon the relationship of philosophy to the study of society, on the one hand, and their relation to social action, on the other. The first of these subjects requires examination in the face of writers like Hughes, who think that the sorts of questions which get raised in Sorel's work, and, more importantly, the confusions they embody, could be resolved on the basis of a more adequate social theory alone. i.e., without rethinking the philosophical assumptions upon which it was based. Thus, it is enough for Hughes to assert that Sorel did poorly, i.e., less lucidly, what Weber, Durkheim and Mannheim did in a more articulate, self-conscious fashion. Even if this is so, there is more to the story. The second of these themes needs to be discussed in the face in the all-but ubiquitous assumption, common to such different thinkers as Marx and Dewey, Habermas and Foucault, that social thought and 'liberation' are inextricably bound. The value of the third arises from the need to articulate an alternative 'scenario' for the relationship of philosophy to social science and social reform.

There are good reasons for beginning this discussion with a number of biographical parallels; for they help to make plausible the value of the sort of inquiry that has just been sketched. To begin with, despite their many contacts and friendships with eminent contemporaries, Sorel and Wittgenstein were isolated figures, outsiders, within the intellectual communities in which they lived. In both cases this was not accidental, but a matter of choice, which was tied to their uncompromising rejection of bourgeois culture. They were equally fascinated by religion and exhibited marked hostility to the pretentions of academic intellectuals. Their works are distinguished for their avoidance of the commonplace. Further, their perspectives on the problems were in both cases decisively moulded by their training as engineers: from the start Sorel and Wittgenstein were less interested in how things looked, or what people said about them, than with how they worked. Conversely, science meant applied science for both of them: idealizations were important only to the extent that they could help to get a job done. It is, therefore, hardly accidental that they were both profoundly impressed by the thought of William James, from whose *Varieties of Religious Experience* Wittgenstein tried to learn to be a saint at one point and to whom Sorel once referred as his philosophical savior. Moreover, no small part of this effort to see Wittgenstein as a corrective to Sorel takes its sense as part of the task of reconsructing the development of Jamesian pragmatism. We forget at our peril that James is, after all, the most cited philosopher in Wittgenstein's *Philosophical Investigations* and that Sorel was the author of *On the Utility of Pragmatism* as well as *Reflections on Violence.* Wittgenstein and Sorel resembled James in being anti-rationalistic without being anti-intellectual. All three aimed, albeit in very different ways, to assess accurately the role and limits of reason in making action possible and, hence, in shaping human destiny. However, for all that, if anything bound Wittgenstein and Sorel together intellectually, it was something that was alien to the spirit that informs James's work: the notion that progress in illusory.

Similarly, in spite of the parallels in their backgrounds and attitudes there was one crucial difference between Wittgenstein and Sorel: Sorel was, as Wittgenstein was not, a *moralist* and an *activist*. Sorel wanted nothing more than to change the world in such a way as to make room for that heroic action that bourgeois society stifled. For Wittgenstein, who was certainly as critical of the vacuousness of bourgeois mores as Sorel, *all* sorts of social prophecy were anathema. "The philosopher", he wrote, "is not a citizen of any community of ideas. That is what makes him into a philosopher".[2] Sorel was convinced, as Wittgenstein was not, that there was a particular type of knowledge which could only be obtained through *participation* in a social movement. Without this experience a part, indeed, the most important part, of social reality remained inaccessible to the social scientist. For Wittgenstein, on the contrary, it was precisely such participation which *prohibited* the participant from getting clear about the basis of his action. This view was congenial to Sorel temperamentally *and* forced upon him conceptually by the Bergsonian metaphysics in whose terms he thought. The problems with respect to vocabulary and conceptual scheme, which Stuart Hughes identified as problematic in Sorel's writings are rooted in those Bergsonian metaphysical commitments. But this is to run ahead. What is clear is that both Sorel's and Wittgenstein's claims here are radical ones inasmuch as they both reject the liberal assumption that the social scientist can be both scientist and reformer. For Sorel, scientific detatchment is incompatible with understanding the forces which change society and, thus with social 'science'. For Wittgenstein, on the contrary, the pre-condition for understanding action is a detatchment as complete as Heideggerian *Gelassenheit*. If either is right, the standard assumptions of social scientists stand in need of revision.

What, then, did Sorel mean by social myth? He defined it in his *Reflections* in terms of three elements.[3] A social myth is (1) an image of a battle, which (2) represents a grand social movement, (3) to the participants in that movement with a view to assuring their victory. Sorel's examples of these models for insuring victory in social conflicts include the early Christian picture of the world, the Protestant Reformation, the French revolution, the national ideal, the Marxist revolution,

DeMaistre's notion of history as the struggle against the secular forces of Satan and the Syndicalist vision of the general strike. Doubtless, he would today include that world view designated variously as Thatcherism or Reaganomics under the rubric of social myth. The point is that, while social mythology is principally exemplified in leftist movements, it is by no means limited to them. The importance of these images of battle is their ability to arouse and sustain unswerving commitment to a course of political action. They motivate us to try to change the world, on the one hand, and articulate moral criteria for self-evaluation, on the other. Thus, social myths provide participants with an identity by virtue of their dual role in educing and justifying action irrespective of its outcome.

The two most important aspects of these myths is that they are *unanalysable* and *unfalsifiable*. Social myths resist the analyst's attempt to reduce them to their constituent elements. Their significance for those committed to them derives precisely from this unitary character. It insures that they must be grasped as they are. They present an uncompromisingly clear picture of reality as a whole, which we must accept. That is to say, they do not allow us the luxury of rejecting them in good faith. Indeed, their unitary character is such that the picture of the world they present must be accepted as all of one piece or not at all. To question a single element in that picture is to reject the whole. They are, in the language of Sir Karl Popper unfalsifiable and, therefore, in the eyes of both Popper and Sorel irrationally maintained.

Indeed, Sorel chose the term myth to describe this mode of belief precisely for its religious connotations. Like early Christians, sixteenth century reformers and Jacobins, modern social activists are a force in the world on account of their unshakable commitment to their cause. In short, revolutionary ideology is better understood as a way of living than as a system of ideas. A system of ideas can be subject to critical scrutiny in a way that a social myth cannot because it is always possible to question the empirical assumptions and logical coherence of that complex. However, Sorel's social myth is only complex on the surface. It is not possible to argue against the revolutionary, the syndicalist or the integralist because their beliefs are not held open to intellectual discussion. They systematically shut out everything that stands in the way of effective action. The latter, then, is based upon just the sort of fanatical belief that philosophers have traditionally taken to be irrational.

In effect, Sorel, an ex-Marxist on the way to becoming a monarchist, was articulating a Bergsonian critique of Marxim. The latter presented itself to the world as having a 'scientific' basis, something Sorel did not deny was in fact present, but it was crucial for him to distinguish that scientific or utopian element in Marxism from its mythic side. For Sorel any program for social reform must have a 'scientific' or utopian component to the extent that it posits a social ideal which is normative with respect to the present state of society and itself subject to criticism. In this Sorel was explicitly Platonic and Neo-Kantian. However, philosophical reflection of a critical sort is woefully inadequate as the basis for social action. For Sorel this was inextricably linked to Bergson's notion that all analysis is based upon static categories, which lay at the bottom of rational demonstration, but which, for that very reason, do not reach deeply enough into our primitive nature to move us to act. But this was little more than a recasting of Bergson's dichotomy between time and duration in such a way as to articulate the difference between the primitive and rational aspects of social movements. It is important to point out that Sorel was very much aware that these two aspects of social thought were rarely found completely separated from one another. They were usually mixed — and for good reason. Social philosophy needs myth to inspire action, but myth too has to have some minimal cogency. Social myth cannot be entirely fanciful; it has to have some contact with reality. Philosophy and myth need each other. But, if this is true, Sorel's own task turns out to be a very tricky one; for it entailed being at once inside and outside the social movements which he would study. He had to be inside them to obtain the participant perspective; outside to make sense of it as a social scientist. It is worth noting that in his own eyes this was little less than reconciling Nietzsche with Kant. Sorel's method of *diremption*, or analysis based upon abstracting cross-sections of social phenomena, is a way of simplifying a reality whose complexity tends to bewilder us is noteworthy for its similarity with Wittgenstein's notion of a language game which serves the same purpose. Given the holistic character of social myths, it was absolutely required methodologically, if Sorel was to produce an analytic account of irrational belief.

Here is where the problem of the limitations of his vocabulary and conceptual framework come to the fore. It will be worthwhile, therefore, to dwell on the sort of problem which confronted him. Sorel wanted to articulate the irrational structure of social action. But to speak of a structure of the passions, which could be identified and analysed, is at least to suggest that there is a 'rational' character to 'irrationality' in some sense or other. Sorel was hung up on a *sharp* distinction between the rational as the scientific or logical, in Bergson's terms what is understood through the static categories of thought, and life, which is spontaneous movement that can only be intuited. The impetus for the individual to identify himself with the movement, to participate in the myth, came from the way in which the categories of logic and a-fortiori science pressed him into the notion that all rational structures are cage-like. If this were the case, then it was only reasonable that primitive instinct rebell against routinization. It is, then, hardly accidental that Sorel turned to the vocabulary of religious experience, that repository of primordial intuition, to articulate what he took to be beyond the capacity of positivistic social science to grasp. There can be little doubt today, after the 'Kuhnian Revolution' in the philosophy of science, that Sorel, like so many of his contemporaries, was pressed into overstating his views about social myth because the whole discussion of rationality at that point in time revolved around determining the exact relationship between the immediacy of pre-conceptual sensory experience, which could only be intuited, and their representation, which was a matter of mathematics. The perspective of practice, i.e., proceeding from what people actually *do*, had not yet made its impact upon metaphysical and epistemological debate. To be sure, there was James, and Sorel was among the first to recognize his importance, but, it must be remembered that James expressed his pragmatism in a psychological mode with the result that the implications of pragmatism for metaphysics and epistemology had to be worked out by others. Thus, for all Sorel's enthusiasm for James, he remained trapped in Bergson's inverted positivism.

At the risk of making a long, complex story all too short and snappy, I want, nevertheless, to insist that the crucial element in our appropriation of James in metaphysics and epistemology came from Ludwig Wittgenstein.[4] Briefly, Wittgenstein's importance for the debates in which Sorel was immersed consisted in his pragmatic recovery of the Hegelian insight that all awareness was mediated by language. By putting language, rather than the dichotomy between intuition and analysis, at the center of philosophical discussion Wittgenstein was able to show us how the idea of pre-conceptual experience was based upon a misunderstanding of the notions of concepts and experience, and, in fact, a contradiction in terms, as well as how it was that concepts were anything but the static cages that they were for Bergson. All of this is stock-in-trade within the Wittgenstein industry. However, the implications of this radical transformation of our understanding of knowledge and action for the sorts of problems which bothered Sorel, and which led him to posit the notion of social myth, have hardly been touched upon. They are nowhere more clearly formulated than in *On Certainty*.

Reading this, Wittgenstein's last work, after having considered Sorel's views about social myths, there is a temptation to be overly impressed either by the differences or the similarities in their thinking. The lack of overtly political themes in Wittgenstein's work might tempt the politically oriented reader to think that they have nothing at all in common; while Wittgenstein's insistence upon the groundless, 'mythological' character of that which grounds true and false propositions might tempt him to think that there is a greater similarity than actually exists. Both views would be very much mistaken: there is neither an identity nor an opposition between these two thinkers but rather a set of resemblances within the family of pragmatists. If Sorel represents an effort to do pragmatist social thought, Wittgenstein is a curious kind of pragmatist metaphysician. To put the matter in a quasi-Kantian manner, if Sorel is concerned with the pragmatic conditions for social action, Wittgenstein is concerned with the conditions for the possibility of action as such. My task is to sketch the significance of this difference.

One of the most striking features of *On Certainty* is — or ought to be — the fact that Wittgenstein's views there are, at least on the surface *more radical than Sorel's*. I refer to assertions such as the following:

(1) the basis for *all* action is *mythic*;
(2) myth is the unfalsifiable ground of truth;
(3) unfalsifiability is not a defect;
(4) a sort of mythology is at the basis of reason-giving, but the latter is not on that account irrational;
(5) it is an error to oppose the primitive to the modern of the sort sometimes termed category mistake;
(6) politics attaches to the incommensurability characteristic of our different mythologies.

Thus, whereas Sorel wants to say that political activism is rooted in uncritical acceptance of a picture of the world which at once motivates and justifies action; Wittgenstein wants to insist that *human action as such, i.e., inclusive of scientific knowledge is rooted in a quasi-mytho-*

logical world picture. Here it is best to let Wittgenstein speak for himself:

I did not get my picture of the world by satisfying myself of its correctness; nor do I have it because I am satisfied of its correctness. No: it is the inherited background against which I distinguish between true and false.

The propositions describing this world-picture might be part of a kind of mythology. And their role is like that of rules of a game; and the game can be learned purely practically, without learning any explicit rules.[5]

Thus, the language games or practices through which we are socialized incorporate a *Weltbild*, which might be described in a quasi-Durkheimian manner as a collective representation of social experience, against which we learn to discriminate between true and false.

Like Pierre Duhem before him and Willard Van Orman Quine after him, Wittgenstein was convinced that all testing of hypotheses was done only within the confines of a system:

All testing, all confirmation and disconfirmation of a hypothesis takes place already within a system. And this system is not a more or less arbitrary and doubtful point of departure for all our arguments: no, it belongs to the essence of what we call an argument. The system is not so much the point of departure, as the element in which arguments have their life. (*OC*, §105)

However, unlike Duhem and Quine, Wittgenstein's point is not confined to philosophy of science, but is a transcendental feature of human experience as such. Thus, *pace* Dewey and Quine, science turns out to be continuous with human life rather than the criterion against which human activities are measured. Wittgenstein was not a conventionalist in the sense that he believed that there was an arbitrary choice between systems of testing; rather, he was calling our attention to the way in which our practices are, as it were, self-grounding:

"An empirical proposition can be *tested*" (we say). But how? and through what? What *counts* as its test? — "But is this an adequate test? And, if so, must it not be recognizable as such in logic?" — As if giving grounds did not come to an end sometime. But the end is not an ungrounded presupposition: it is an ungrounded way of acting. (*OC*. §§109—10)

What we learn is a whole system of judgments, which together constitute an unshakable picture of the way the world is. It is important to emphasize that for Wittgenstein this is a *fact of nature* — one whose obviousness and generality inclines us to overlook its importance for dissolving the problems traditional philosophers pose. In any case, we do not learn this quasi-mythology as an explicit picture of the world but in the course of learning to make judgments. Wittgenstein compares the language games or practices through which this nest of judging techniques is mastered with a game that can be learned entirely through imitation. Thus, he insists that all knowledge is based upon experience — though not on *my* experience; rather on the collective experience in terms of which I am socialized (*OC*, §275).

This point can be reformulated to assert that the basis of knowledge is trust: "I really want to say that a language game is only possible if one trusts something". (*OC*, §509) For Wittgenstein the sorts of expectations that we have about the world, i.e., what constitutes our logical grammar, are rooted, not in explanations, but in what we learn about the world. (*OC*, §286) This is what Wittgenstein means when he says that explanation must come to a stop: it stops with certain basic facts about the world, which we accept on trust. These facts are "hardened" into forming the basis of practice. (*OC*, §96) Their "hardening" is not a matter of their having a peculiar character or set of intrinsic properties such as, say, the Kantian synthetic a-priori, but that they come to play the sort of role Kant prescribed for synthetic a-priori propositions. This is not some arbitrary practice in the sense that any given individual decides what is to stand fast. (*OC*, §271)

To say that what stands fast for us is not subject to falsification is to say that certain assumptions are the basis of action, that our practices depend upon them, and that calling them into question, as philosophers try to, will seem silly or unintelligible:

> I am sitting with a philosopher in the garden; he says again and again "I know that that's a tree", pointing to a tree that is near us. Someone else arrives and hears this, and I tell him: this fellow isn't insane. We are only doing philosophy." (*OC*, §467)

It is a cardinal tenet of Wittgenstein's views in *On Certainty* that certain types of mistakes, i.e., those relating to our *Weltbild* or mythology, are difficult to distinguish from madness. (*OC*, §73) Thus, it is not that failure to question what is presupposed in the picture of the world in terms of which we have been enculturated is a matter of stupidity or credulity (*OC*, §235), our trust in them is what makes us able to get around in the world. Interestingly, on this account, the justification for the kind of silliness or insanity which attaches to philosophy rests upon the fact that these webs of belief occasionally cease to be reliable. Then we no longer know our way around and need to discover why.

It is critically important to reiterate here that Wittgenstein is not defending a relativism according to which individuals arbitrarily decide what constitutes a *Weltbild.* He insists that there is a crucial difference between something seeming to be so and actually being so at the very beginning of *On Certainty.* (*OC*, §2) The world picture in question is implicit in everything we learn, rather than being articulated. One of the least appreciated but most important implications of this view, as I have suggested elsewhere, is the potentially political character of our picture of the world. This is, of course, something Wittgenstein never explicitly discussed; yet, there is a passage in *On Certainty* which, at least on the face of it, lends support to this hypothesis:

> Suppose we met people who did not regard that [allowing ourselves to be guided by the propositions of physics] as a telling reason. Now, how do we imagine this? Instead of the physicist, they consult an oracle. (And for that we consider them primitive.) Is it wrong for them to consult an oracle and be guided by it? — If we call this 'wrong' are'nt we using our language game as a base from which to *combat* theirs? (*OC*, p. 609)

There seems little doubt, then, that Wittgenstein saw certain types of conflicts as rooted in the mythology of our world pictures. To be sure, this is something different from the sorts of explicit images of battle which characterize Sorel's social myths. Nevertheless, it seems clear that Wittgenstein is committed to the notion that at least some political clashes are part and parcel of our natural history.

But what should we learn from all of this? If Wittgenstein is right in these matters, the most central consequence for Sorel's views about social myth is that what the latter took to be irrational, because ungrounded or illogical, is in fact normal. Here we have to be careful, for it is possible to read a kind of irrationalism into *On Certainty*, which is in fact not there. When Wittgenstein says, for example, "you must bear in mind that the language game is so to say something unpredictable. I mean it is not based upon grounds. It is not reasonable," (*OC*, §559), we have to bear in mind that he adds "(or unreasonable)". It is there like our life". Here Wittgenstein means to say that language games or practices are the distinguishing features of the kind of animal that we are, that they identify our place in nature, and, thus are neither more nor less rational than nature herself. Here, it should be pointed out, Wittgenstein is, deliberately or not, close indeed to Jamesian pragmatism, whose aim was to explore the implications of Darwin's notion of natural selection for philosophy. This was not a matter of producing an evolutionary theory of knowledge and mind, as say, the German

Monists would, but of seeing how the wide variety of activities that were human were all rooted in nature. For James pragmatism could just as well be called pluralism. In any case, the point was to see the continuity between nature and mind. Wittgenstein's language games can legitimately be seen as carrying out a Jamesian task here. The point of this undertaking was to show that the unquestioned basis of action — and language — is neither rational nor irrational.

From a Wittgensteinian perspective human practices which endure over time are rational at least in the sense that they incorporate reasoning — even if they are not based upon the practice of giving reasons. Much of what is problematic about Sorel's thinking is tied to the narrowness of his concept of rationality. On his view there was a once and for all absolute dichotomy between the rational and the irrational, which was based upon equating the rational with the formal sciences of logic and mathematics. Thus, he could hardly hope to escape having to term everything that did not have a logico-mathematical character irrational and, therefore, 'primitive' and instinctual. It is the merit of Wittgenstein's thinking on these issues to dissolve the difference between the 'modern' and the 'primitive', between the rational as a specific set of 'scientific' beliefs and what is not 'scientific', in the end, between mathematics and magic, in such a way that we are able to see the rationality of the primitive and the 'primitive' character of the most scientific reasoning *both* as rooted in analogous sets of unquestioned assumptions. Wittgenstein's greatness consists precisely in seeing what others, Sorel included, failed to see because he did not identify rationality with any specific set of beliefs but with the way that models, hypotheses and assumptions functioned in human life.

If Sorel chose to describe the sorts of representations of conflict that moved men to act as *myths*, it was because they had the power to make us move without reflecting, it was because he was convinced that their power was tied to something 'primitive' in the sense of being demonic, which reason could not reach and which social convention masked. If Wittgenstein characterized the propositions which would adequately describe a practice more cautiously as a "kind" (*Art*) of mythology, it was just to stress the *problem* we have of representing to ourselves those pictures of the world which make action possible. Sorel's view was full of problems methodologically. In principle the actor could not analyse and the scientist could not act without both ceasing to be what they are. From a Wittgensteinian perspective there is something to this idea but it is not because there are two radically different modes or faculties for knowing, rather it was a matter of the difference between knowing how to do something and understanding what we have done.

Wittgenstein renounced changing the world for the sake of understanding it precisely because acting and understanding, the basis of our action, are not compatible with one another. The sort of understanding he sought was not merely an understanding of any specific action but of action itself. To understand action as such was a matter of grasping general features of human nature relating to what it is to follow a rule, which were paradoxically obscure because they were universal and obvious to all but the theorist. Thus, he had to concoct an elaborate procedure of self-questioning, which called the standard assumptions about knowledge into question in such a way as to show that seeing how things hang together was an intrinsic element in the logical grammar of action. The result was a praxis-immanent account of rationality, which was capable of accounting for why it was so difficult for us to give an account of what we know. If to know was to be able to *do* something, it follows that knowing how to explain what we know is at best a secondary matter. We learn to do by following a rule *blindly*, and this explains why we have so much more trouble explaning what we do than in just acting. Dancing, football and swimming are three good examples here.

In *showing* us this through his questions, examples aphorisms and the like Wittgenstein showed as well what was wrong with the sort of Bergsonian assumptions that Sorel had made about rationality. It would seem that the price he paid for this insight was too high; for having understood the 'mythic', unfalsifiable and, if I am not mistaken the political character of action as such, it was necessary to renounce influencing the course of events. Moreover, we seem to have lost focus upon the most direct goal of Sorel's inquiry, determining just what makes people into political activists in the normal sense of the term politics. There is, again, certainly some truth in those allegations but that should not obscure the fact that Wittgenstein does indeed have something important to say to somebody with Sorel's concerns. Let me deal briefly with these questions in reverse order.

From the Wittgensteinian point of view there seems to be good reason to maintain that the study of politics be extended to every sphere of activity inasmuch as it is based upon questionable, yet unquestioned assumptions, but now the center of the study of politics will be changes of perspective involving the unquestioned foundations of our enterprizes from poetry to physics and from architecture to musical taste. In short, the study of politics becomes the comparative study of shifts of perspective of a wide variety without being restricted to what is normally understood as public life. This is important because it 'humanizes' politics in the sense that it establishes that shifts of perspective are as such political, i.e., that what Sorel *had* to assume was something irrational, was part and parcel of human action as such, and therefore, a 'normal' part of human life which we should expect to encounter in all areas of experience.

The other problem is more difficult to deal with. Wittgenstein's insistence that the philosopher's lack of membership in a community is his defining characteristic is as puzzling as it is uncompromising. However, if it is the sort of philosophical bedrock which Wittgenstein normally sought, he should be understood as maintaining that, however much a philosopher might want to participate in the life of a community, he can never succeed in doing so. This is because to succeed in being involved in a community of ideas is to cease to be a philosopher. Thus, he does not mean that people calling themselves philosophers will never act as members of such a community but that in so acting they cease to be philosophers. If by any chance he should turn out to be right about this matter most of the people who have been pronouncing upon this matter, including people like Habermas and Foucault as well as Sorel and Dewey, turn out to be very mistaken. Therein lies the very fascination of his view and the challenge with which he presents us.

NOTES

[1] H. Stuart Hughes, *Consciousness and Society* (New York, 1958), p. 181.

[2] Ludwig Wittgenstein, *Zettel*, trans. G. E. M. Anscombe (Oxford, 1967), # 465.

[3] Georges Sorel, *Réflexions sur la violence* (Paris, 1981), p. 26. My account of Sorel's notion of social myth is drawn from his introduction, pp. 26—42 and from the fourth chapter, "La grève prolétarienne", pp. 49—56. I have drawn heavily upon Stuart Hughes in discussing the more general features of Sorel's conceptual scheme.

[4] I have examined Wittgenstein's relationship to William James in unpublished papers treating their respective accounts of mysticism and the pragmatism of the *Tractatus*.

[5] Ludwig Wittgenstein, *On Certainty*, trans. G. E. M. Anscombe and Denis Paul (New York, 1972), # 94—5. For convenience I shall cite this work parenthetically in the text by means of paragraph numbers.

CHAPTER XI

SELF-DECEPTION, NATURALISM AND CERTAINTY: PROLEGOMENA TO A CRITICAL HERMENEUTICS

In August Strindberg's Preface to *Miss Julie* there is a statement of what literary Naturalism took to be its new concept of motivation:

> What will offend simple minds is that my plot is not simple, nor is its point of view single. In real life an action — this, by the way, is a somewhat new discovery — is generally caused by a whole series of motives, more or less fundamental, but as a rule the spectator chooses just one of these — the one which he can most easily grasp or the one that does most credit to his intelligence.[1]

In the wake of this "overdetermination" according to which there can be multiple accurate accounts of what an action is all about philosophical Naturalists have inferred that first person accounts of behavior are notoriously unreliable.[2] Moreover, they insist that our tendency to produced clichéd or flattering self-descriptions, ones which fly in the face of psychological experiments and psychotherapeutic experience, warrants rejecting hermeneutics *tout court* as a source of knowledge about human action. Self-deception, they argue, is the normal, perhaps inevitable, result of overdetermination. Because that is so, Naturalists insist that the only adequate account of human agency must be a theoretical explanation of behavior rather than an interpretation of action. This was not Wittgenstein's view, but there are so many curious similarities and differences between the two to insure that a Wittgensteinian critique of the Naturalist account will contribute to everybody's understanding of the problem. The aim of this exercise, then, will be to pursue such a critique with a view to vindicating a hermeneutic perspective on action, but one which, precisely on account of its sensitivity to the problem of self-deception and accurate self-description, can justly be termed a *critical* hermeneutics.

What sorts of claims can be made on behalf of over-determination and the attendant self-deception with respect to our own action that it would appear to entail? Three sorts of claims can be advanced. The first is the strong claim that all accounts of behavior entail self-deception. It requires no discussion because it is absurd. The second is the weak claim that some accounts of behavior, such as, say, that of neurotics and psychotics, are self-deceived. This is an uninteresting claim; for it does not seriously challenge what most people believe. It is the third claim, namely, that most first person accounts of behavior are self-deceived, with its corollary that hermeneutics is simply a mode of participating systematically in that self-deception and its implication that only a science of the mind will permit us to explain the way things actually are, which is philosophically interesting.

On what basis is such a claim advanced? Two sorts of evidence have been marshalled on its behalf; psychological experiments and the successes of psychotherapy.

Experimental psychologists have carried out various sets of experiments designed to show that subjects cannot possibly be correct in their own accounts of how they go about solving problems.[3] Typically (and somewhat over-simplifiedly) subjects are asked to memorize a list of, say, six numbers flashed upon a screen, then to pull one or another lever depending upon whether another number subsequently flashed upon the screen was or was not on the original list. The experiment in question found that the reaction times of the subjects did not correlate with either of the two explanations which subjects gave of how they found the number on the list. They concluded that we do not know how we perform such tasks.

But what do such results actually imply? Clearly, Naturalists rightly infer on that basis that we are not "omniscent" with respect to our mental activities but they tend to think, reinforced by a dose of Skinnerian metaphysics,[4] that such results confirm the controversial thesis that we are generally unable to give an accurate account of what we know, that we need a science of the mind to get to the bottom of the matter. Yet, there seems to be another view of the matter, one which would proceed from the very evolutionary perspective so dear to the Naturalist. It is possible to ask precisely in the case of such experiments as the one described: do they test skills that actually play a role in our struggle for survival? On the face of it they do not. But, if this is right, there is no reason whatsoever to worry about our inability to give a reliable account of what we do. On this view such 'experiments' are closer to a variety of puzzles or even to tricks played upon us than they are to anything that might count as a normal mental process.[5] In short, there is no question that such experiments indicate that we are by no means "omniscent", with respect to our own mental activities, but it is mighty unclear what we are to infer from that. Such experiments have a sense as long as we conceive of psychology atomistically, i.e., as discovering the building blocks of mind. If we do not so conceive psychology, they are, for the moment at least, mere curiosities.

The successes of psychoanalysis present us with a very different matter; for there the case for self-deception is much more compelling.[6] It is the undeniable strength of Freudian therapy to have demonstrated to us that many, if not most, of the mental problems we encounter are linked to repression, which entails self-deception. The successes of Freudian therapy are undeniably based upon the analyst's ability to convince the neurotic patient that his normal way of describing his motives involves systematically hiding his actual motivation from himself. Said successes are certainly strong evidence for the kind of psychological damage self-deception may cause. However, the crucial question here is: do we need a *theory* to unmask self-deception? The sense in which the Freudian therapist *needs* a theory is interesting indeed. Unlike the family physician, whose cures work or not, regardless of what patients think about them, it is a necessary condition for successful analysis that the patient accept Freudian theory. From the point of view of philosophy of science, however, this is just what makes the theory suspect. So, what starts out looking like grist for the Naturalist's mill turns out to be hardly that. Alasdair MacIntyre has brilliantly and painstakingly demonstrated how the successes of Freudian therapy need not be taken as evidence in favor of the theory.[7] His point is that the successes of psychotherapy can be accounted for in terms of the way it makes a new scenario for living plausible for the patient, which is in essence to claim that its successes are the result of its hermeneutic properties and the rhetorical gifts of the therapist, whose function it is to persuade the patient of the truth of the Oedipal situation as the actual scenario in terms of which he has lived his life.

The point of all of this is that on such a view the successes of psychoanalysis present us with a case in which we substitute autophenomenology for heterophenomenology, to use the terms favored by the Naturalist,[8] but there is no reason for insisting that the therapist be armed with a scientific theory. In fact, *persuasion* and *trust* turn out to be the crucial elements in a successful analysis. (We forget at our peril that Freud was wont to suggest that patients with highly developed super-egos were better off to see their priests[9]). What we should learn from this is that individuals may have problems with self-deception, which can be overcome on the basis of the community's collective understanding of its own practices (I take this to be one of the reasons why Aristotle insisted that moral philosophy could only be pursued communally[10]).

But what has all of this to do with Wittgenstein? Quite a lot. For starters, his *Lectures and Conversations* suggests that he endorsed just the sort of critique that MacIntyre developed. Brian McGuinness has made a strong case for the claim that Wittgenstein's concept of philosophy in his later work was strongly under the influence of Freud.[11] McGuinness argues persuasively that Wittgenstein's notion of philosophy as therapeutic was directly inspired by his critical encounter with Freud's practices. Moreover, *On Certainty* treats the problem of misrepresenting our own beliefs centrally and in a way that illuminates the sort of considerations which Naturalists want to discuss under the rubric of self-deception.

In his *Lectures and Conversations on Aesthetics, Psychology and Religious Belief* Wittgenstein makes what to his way of thinking is a devastating criticism of psychoanalysis. "Freud", he is reported as saying, "never shows us how we know when to stop — where is the right solution".[12] It should be remembered that in his later period he as wont to characterize the aim of philosophy as reaching a certain kind of "bedrock" (*Fundament*) (*OC*, §498) where the desire for explanation comes to an end (*PI*, §133). His assertion, then, is that Freudian therapy brings the interpretive procedure to a halt arbitrarily. He takes it as evidence for that arbitrariness that Freud's criterion for correct interpretation can shift from (1) one that the patient accepts to (2) one that the patient does not accept but that the doctor 'knows' on the basis of his theory to be 'correct', but which the patient 'resists'. It is important to emphasize that Wittgenstein, unlike many other critics of Freud such as Laing or Szasz, in no way intends to belittle Freud; indeed, McGuinness reports that he went so far as to proclaim himself a "disciple" of Freud. For Wittgenstein the important thing was to set Freud into perspective. He turns the tables on Freud at the crucial point that when Freud insists that we are disinclined to accept Oedipal accounts of our motivation by insisting that "if the explanation is one which people are *dis*-inclined to accept, it is highly probable that it is also one which they are *inclined* to accept" (*L&C*, p. 43). He thinks that there is a certain pre-conscious acquiescence in Freudian explanation which disposes us to be persuaded by the therapist. No less important is the fact that he rejects Freud's account of self-delusion because he rejects the notion that there can be one true account of the basis of belief which excludes all others. To see his path to this radical view we must look closely into his views about the nature of knowing, certainty, mistakes and madness as they are developed in *On Certainty.*

We can formulate the main points that Wittgenstein wants to make apropos themes relevant to the Naturalist's problems stemming from self-deception:

(1) Individuals can often, *but not always*, be mistaken about *what* they know. Their accounts of *how* they do what they do need not be correct.
(2) The groups through which individuals are socialized are always *potentially* in a position to provide an incorrigible and correct interpretation of the order in their practices to the extent that those practices are stable and homogeneous.
(3) Scientists are not in a privileged position with respect to understanding our practices.
(4) This is because language is so related to the world that it disguises just the things that we would most like it to reveal. Thus, our interpretations have to be critical in the sense that their success depends upon developing a critical stance with respect to the way language tempts us to see our practices wrongly.
(5) *Nobody* is ever in a position in which they must change their interpretations.
(6) Interpretation is intrinsically *political* to the extent that practices can be construed in differing ways.

These statements require elaboration.

Wittgenstein considers that human behavior is always rooted in a picture of the world that he likens to a myth:

> The propositions describing this world-picture might be part of a kind of mythology. And their role is like that of rules of a game; and the game can be learned purely practically, without learning any explicit rules. (*OC*, §95)

it is against the background of this picture that we learn to discriminate between the true and the false but it is important to emphasize that this picture is implicit in action. Here there is much to be said for, distinguishing at least two significantly different senses of the word 'know' (which is an absolutely central part of Wittgenstein's task in *On Certainty*) and reflecting upon the implications of such a distinction.

To 'know' in the most basic sense for Wittgenstein is to be able to *do* something; whereas to know in the sense of being able to give an account of a given phenomenon is something quite different. The western philosophical tradition got off to a bad start in discussing these matters when Socrates insisted that they were one and the same in rejecting the artisans' claims to knowledge on the basis of their inability to give an account of themselves. For Wittgenstein there is no reason at all why we ought to be able to explain what we know. His pragmatism, then, leads him to reject first person accounts of knowledge. Understanding how people think, what it is to know etc. is a matter of describing the quasi-mythic *Weltbild* implicit in their actions — a hermeneutic task, but, a critical one.

Describing these world pictures is a matter of getting straight about the "grammar" or "logic" which makes our practices coherent — of learning how we weave together words and actions. It is a matter of learning to determine the meaning of expressions by looking at the way we use them. It does not depend upon taking agents' self descriptions as the last word; for the point of words is to get a job done, not to inform others accurately of exactly what we believe. Knowledge in the most basic sense is a matter of learning to do something. We know when we have succeeded in performing the requisite operation. Our accounts of what we do, however, stem principally from those who are trying to train us to perform the required act, say, to serve a tennis-ball properly, and, consequently, have more to do with *building* confidence, which entails establishing a certain self-image in us, and *preventing* us from becoming distracted, i.e. with giving us the proper orientation towards our task, than it does with informing us about the principles upon which the instruction is based. In teaching someone to play tennis the coach does not explain what he is doing but simply *orders* the would-be tennis-player to pay attention to and imitate certain examples and to avoid certain temptations. There is no room for a "why?" in the instruction because the answer is always obvious: "that's just how you learn!". The aim here is to know something 'with the body' rather than 'with the mind'.

This instruction has a specific 'natural history', i.e., internal ordering — logical grammar — which fits the kind of agent-knower that we happen to be. It begins with our need to be *drilled* and proceeds through a point a which the behavior patterns, which are *caused*, come to be understood in terms of *reasons*, to a point at which our unreflective efforts at articulation tend to become entangled in confusions about the causes of and reasons for acting the way we do.[13] We shall see that language plays a crucial role in confusing us.

To play a game such as tennis we must learn to respond to specific commands. The latter play a causal role in establishing a routine for accomplishing a specific task. Understanding how we do what we must is not important here — and the demand to know counter-productive. Getting the job done is everything. Things get complicated because the language in which we are instructed is not intended as a vehicle for understanding the character of the practice but for initiation into it. Thus, skill in performing an action entails knowing how to talk about it, but that by no means implies an *understanding* of the order in the practice; for the language we have learned was part of learning the game; not how to articulate its structure. For this task we require the sort of *reflective experience* typical of coaches.

Perhaps the best way to grasp this is to reflect for a moment upon Wittgenstein's assertion, ". . . it has no meaning to say that a game has always been played wrong". (*OC*, §496) This is one of his famous "grammatical" points: if a game has always been played wrong; it has not been played at all; we have been doing something else. It is part of our understanding of games that the people who play them know what they are doing. This follows from the nature of games as practices, i.e., rule-governed activities, that they have to be mastered through repetition over time: because rule-following activity is a 'hit or miss' activity, we have to be trained to 'hit' more than we 'miss'. For this reason in games — and in life — we need specialists in training, who are in fact people whose function it is to teach us the kind of skills or competences we require to be able to master the practice. They teach us how to be certain in action, as it were. They also correct us when we are off-form (note the similarity with the role of the psychotherapist as conceived by MacIntyre).

When we want to understand what we do, it is necessary to reconstruct our experience reflectively. This is what coaches do and their role is one from which we can learn a great deal about the hermeneutics of action. The coach is a person with an *experienced* third person perspective on a practice, which is to say that the coach both knows how to play the sport in question and how to convey a sense of the fine points of the game to players. To know the game, then, is to know what a competent coach knows; for that knowledge is definitive with respect to the practice. The suggestion, then, is that Pragmatists who want to learn what internal realism is all about have a lot to learn from coaches and coaching. A hermeneutics of action reconstructs a coach's knowledge of a practice.

This example is informative because it casts light upon the concept of expertise as Wittgenstein would have us understand it, of how it is that what ordinary agents 'know' can be obscure to them (in the manner of a myth) without, for all that, requiring a scientific theory by way of explanation; for it shows how agents *must* be unclear about the basis of their action but that they can become clear about it. It also explains why it is that a scientist is in no better position than the agent — and in a worse one than the coach — in understanding the nature of practices.

The whole point of the philosophical enterprize according to Wittgenstein is to put the identity which exists between thinking and acting into perspective. This is necessary because language stands between us and an accurate understanding of our own action. Both scientists and ordinary agents require a critique of language if they are to see their own practices rightly. Positivism recognized that there was a problem of reference but it conceived this problem differently from Wittgenstein. For the Positivists, as for their Naturalist successors, the chief problem with language is our tendency to think that there must be an object corresponding to a common noun whenever it is possible to use said noun in meaningful sentences. The problem that they find central to their critique of language, then, is the problem of reification. Since the word horse refers to a class of horses, the word unicorn must refer similarly to a class of unicorns — or so it would seem. Positivists took it to be a principle task of philosophy to provide an account of reference which would enable us to demarcate between referring and nonreferring terms. In this they followed Frege, Russell and Wittgenstein's own early work. However, in Wittgenstein's later philosophy the main problem with language is a related, but, nevertheless, essentially different one. It is the temptation to think that there is always some *one* thing to which a word refers when in fact there is no special reason to think that words have to have an unambiguous meaning. It is important to emphasize that this is simply a fact, albeit one of special importance for philosophy.

Perhaps Wittgenstein's point can best be illustrated by looking at the implications it has for understanding our emotions. For the Positivists the program to eliminate reification had as its correlate a program for identifying the observable referent of psychological words like love, fear and joy.[14] In fact this maneuver backfired. It got us out of the Platonic World of Ideas at the price of a certain cultural imperialism with respect to the world of experience. Their problem became to identify a *specific* type of behavior with the word 'love'. This entailed correcting ordinary usage and a concomitant theoretical justification for that correction. In effect, their essentialism led them to a curious kind of foundationalism but one in which scientific theory took the place of philosophical theory.

Wittgenstein's response to the Positivist search for the behavioral constellation that corresponds to a 'mental state' such as love is simply to deny that there is a class of actions that can be so demarcated a-priori. To understand such concepts we have to take a *look* at what people *do* when they love. Wittgenstein invites us to take note that there is in fact no set of characteristics common to all instances where we would speak of acts of love, love as motivation or being in love (cf. *Z*, §§468–94). Doing so, we come to see that love is much more a way of behaving, i.e., 'lovingly', which can modify a wide range of acts, rather than a specific action or set of actions like 'tennis'. Thus, for Wittgenstein such 'mental states' turn out to be as 'physical' and public as the sorts of referents the Positivists were seeking for them but paradoxically they do not turn out to be directly observable because the sort of modification of behavior they represent can only be observed by persons who know what to look for. On this view knowing what love is requires reflective mastery of social practice — or, in a word, experience — of a sort that defies textbook presentation; for it is principally a matter of making a lot of subtle distinctions about the nature of situations, our perceptions of them and our modes of reacting.

Whereas the Positivists and their Naturalist successors were claiming to represent an evolutionary perspective by supplementing a fairly traditional nominalism with psychological behaviorism; Wittgenstein developed a genuinely evolutionary perspective on behavior wholly in the spirit of that pluralism which William James identified with Pragmatism. For Wittgenstein, as for Darwin, the given, as it were, is variation; whereas the task is to give an account of its development in terms of a scheme which allows for all the richness of variety.[15] For Darwin this was natural selection; for Wittgenstein it was the sort of rule-following appropriate to a *speaking animal* — we forget at our peril how much he emphasizes the animal origins of knowing and acting in his last work: "I want to regard man here as an animal; as a primitive being to which one grants instinct but not ratiocination. As a creature in a primitive state" (*OC*, §475). Contrary to the German Idealists, Wittgenstein does not set Nature and Culture over against one another; rather he wants to see how the latter is continuous with the former: how our natural history contains the explanation of the conditions of possibility for knowing and acting.

The point to be emphasized here is Wittgenstein's recognition of the fundamental appropriateness of Aristotle's definition of a human being as a speaking beast (*zōon lógon échon*) — curiously, or maybe not, without ever having read Aristotle — which implies a continuity with everything animal (for Aristotle there is significant comparision with bees and ants) but, nevertheless, a certain distinctive mode of animality following upon the capacity to speak. The difference which this difference makes has hardly ever been more succinctly and trenchantly captured in the way that it was by Aldous Huxley, himself scion of one of the houses in which evolution was discovered and from which it was popularized. In one of his lesser known works, *After Many a Summer*, Huxley writes:

> ... and where there were words, there, of necessity, were memories and wishes, judgments and imaginations. There, inevitably, were the past and the future, the actual and the fantastic; regret and anticipation; good and evil; the creditable and the discreditable; the beautiful and the ugly. Among men and women, even the most apparently bestial acts of eroticism were associated with some or all of these non-animal factors — factors which were injected into every human situation by the existence of language.[16]

This is in fact Huxley's reply to Lawrence, but it could well have been Wittgenstein's reply to Quine with respect to the continuity between the human and the mute-animal.

In any case, the complexity of the link between language and the world is such that both scientists and ordinary mortals shall always be tempted to seek the wrong sort of answers, i.e., theories, for the questions that puzzle them most, whose actual resolution rests in a clear grasp of a few "general facts of nature" (*PI*, II, xii, p. 230). Philosophy is, therefore, pragmatically allotted the role of that sort of conceptual analysis which unties the knots that the metaphysician within all of us ties himself up in. It is an *activity* which can be thought of as above or below science but is never on the same level with it — which is not to say that it is in any way opposed to science. In fact, it is the precondition for seeing our scientific problems rightly. It is important to point out here that the thorny question of whether such an activity can be institutionalized has received all too little attention by Wittgensteinians.

To return to the main argument. The quasi-mythic foundation of thinking has a consequence that we ought not to be expected to be in a position to give an accurate account of how we do what we do; however for Wittgenstein that is far from being the whole story, since it is absolutely central to the position that he develops in *On Certainty* that we are by no means on that account alone to consider ourselves as self-deceived. Indeed, it is one of the most radical claims in that little book that we can believe what is false legitimately. Certain belief — which is yet another sense in which we use the word 'know' in the first person — is sharply distinguished from true belief:

> In certain circumstances a man cannot make a *mistake*. ('Can' is here used logically, and the proposition does not mean that a man cannot say anything false in those circumstances). (*OC*, §155)

> In order to make a mistake, a man must already judge in conformity with mankind. (*OC*, §156)

As an example of the sort of situation in which Wittgenstein would say we cannot be in a position to make a mistake we are offered the case of a person who keeps on making a mistake about where he lives. In the natural history of human life this is ruled out, not because it never happens, but because this is not treated as a mistake but as a mental disorder. Here Wittgenstein does a remarkable minuet upon Freud using the comparison between mistakes and mental disturbances to point to the very different roles that certain 'empirical' statements come to play in our lives. (*OC*, §§67—73). Normally a person cannot make a mistake about where he lives because this fact is absolutely basic to his knowing/doing anything else. Everything that he knows/does will depend upon that. If that is wrong, then everything he thinks/does is also wrong and that is nothing else than a form of madness. Sanity, then, entails certainty in life.

"What we believe depends upon what we learn" (*OC*, §285), Wittgenstein writes. What we learn depends upon taking certain parts of experience (which could be different) as beyond doubt. In this Wittgenstein is allowing ordinary agents a luxury which Naturalists only allow to scientists; and in so-doing he is avoiding what seems to be a paradox in Naturalistic thought. On the Naturalist view, ordinary agents must always submit their accounts of their behavior to the scrutiny of scientists. They ought to be prepared to accept scientific revisions of their accounts of what they do. Scientists, on the other hand, *never* have to give up their theories. How is it possible to justify their elitist status. Naturalists would say that the privilege accorded to scientists here follows from the fact that their enterprizes are more successful than those of ordinary agents. However, this is not altogether certain. That move only works, if we are permitted to identify science with first-class science. However, it is wholly unclear how we are to do that except with the wisdom of hindsight. The history and sociology of science is replete with evidence for the thesis that scientists behave pretty well like normal members of the communities in which they live.[17] For that reason science is, by and large as prone to fads and fancies as the 'secular' communities in which it is embedded. Moreover, it is that very unfalsifiability entailed by the Naturalist's commitment to holism which makes it impossible, except in the light of history to make the crucial distinction between first-class science and whatever is not first-class. If this is right, i.e., if we are no different from scientists, if science is not privileged with respect to understanding our behavior; then we are not compelled any more than scientists are with respect to theories to abandon our accounts of ourselves when they come under criticism. This is Wittgenstein's view of the matter. It has most interesting consequences when we turn back to, say, consider the kinds of things that happen in the course of psychotherapy.

The most dramatic consequence of Wittgenstein's view with respect to the unprivileged character of scientific accounts of behavior is that, despite our occasional inability to render an accurate account of our own belief system, there is a sense in which our certainty with respect to what we know about ourselves is *unassailable*:

> Isn't the question this: "what if you had to change your opinion even on these most fundamental things?" And the answer seems to me to be: "you don't *have* to change it. That is just what their being 'fundamental' is" (*OC*, §512)

If we reflect for a moment upon the sorts of criticisms that Laing and Szasz bring against psychoanalysis, we find Wittgenstein giving them metaphysical aid and comfort here: ". . . I find it quite correct for someone to say 'nonsense' and so brush aside the attempt to confuse him with doubts at bedrock . . ." (*OC*, §498). Since our identity is tied to what we learn in the course of acting about ourselves, i.e., learning our role in a social myth, it is possible that we can, in the course of conflicts over just what is to be taken for granted come to regard those who represent competing views of "what stands fast in experience" (*OC*, §125) not only as political foes (*OC*, §611) but as mentally incompetent. Wittgenstein is, therefore, far less removed, say, from feminists like Nancy Mairs, who link mental illness in women to "what stands fast" in patriarchal society as those women come to an awareness of an alternative to that frame of reference.[18] It is perhaps the most radical consequence of Wittgenstein's philosophical psychology in *On Certainty* that there is a definite suppressed political element in psychoanalysis — and Naturalism. Not knowing when to stop, i.e., the tendency to go back beyond the beginning is far from harmless. Thus, the question "*what* is to be tested by *what?*" (*OC*, §125) is a meeting point for Wittgenstein and Foucault.

It must be emphasized that, contrary to what sometimes seems to be the case, Wittgenstein makes these political conflicts (now using the term in a wide Aristotelian sense rather than a narrow Hobbesian one) revolve around social groups rather than individuals. This is important lest we fall into the error either of making such conflicts relative to an *individual's* decision or the notion that learning to judge "in conformity with mankind" (*OC*, §156) precludes such strife. True, Wittgenstein does occasionally speak in both idioms, but it is clear, "what is a telling ground for something is not anything *I* decide (*OC*, §271), and "judging in conformity with mankind" does not preclude a situation in which fighting over what stands fast in experience, i.e., the "system of judgments" or "frame of reference" against which I judge is challenged (*OC*, §611).

We are now at the point where we can draw this rambling, diffuse discussion together by drawing its implications for hermeneutics.

To begin with, it is clear that an interpretation of behavior with an aspiration to an adequate description of how we do what we do cannot be based upon first person accounts of how skills are performed (althought first person accounts of intentions have to be taken into consideration, since we are not in a position to assert that somebody's account of what they do is mistaken till we know just what it is that they think they are up to). First person assurances of certainty tell us nothing about the truth of the belief in question in matters in which it is possible to be mistaken. Similarly, knowing the truth about a given phenomenon does not imply that I can say what someone else 'knows for certain' with respect to the subject. First person beliefs are absolutely certain only when denying them would indicate mental disturbance on the part of that person.

Secondly, understanding practices implies a grasp of rules and mistakes such that it is possible to recognize that an agent has 'made the right move' or proceeded correctly. It can not be private or idiosyncratic with respect to the rules; for then we could never check up either on ourselves or on others: ". . . it isn't for example just *my* experience, but other people's, that I get knowledge from" (*OC*, §275). Those practices which we best understand in the sense of being able to give a good account of are those which are at once most public and most stable: the law, sports, monasticism, military drill. For these skills there are experts, who are people practiced in the rules, such as coaches, novice masters, drill sergeants etc. who have mastered the rules in the sense not only of knowing how to follow them but of being able to initiate others into their practice. Understanding 'theories' in terms of practices, as a Pragmatist must, is to interpret practices from the perspective of the coach — who might be a paradigm case of what Hilary Putnam refers to as an 'internal realist'. This is to have a virtual or implicit knowledge of what it is to understand the rules that constitute a particular practice.

Thirdly, the 'Coaching Perspective', if I can call it that, breaks down when there are competing criteria for understanding a practice, when there is more than one plausible way of playing the game (and a-fortiori of describing it). In those circumstances deciding what is at stake (what should stand fast) is a *political* matter. Interpretation is thus seen to be a political activity as soon as we consider the question of competition between varying interpretations; for then it becomes clear that there is no interpretation without authority. From this point of view it should be clear why we often have problems distinguishing madness, sanctity and political revolution as the cases of St. Francis, Nancy Mairs and Moamer Kadaffi indicate. Indeed, it is all but impossible to become involved in the politics of description, when we become involved in the description of action.

Fourthly, the Hermeneutics of mind, which is to displace the Science of the Mind — without, on that account in any sense being anti-scientific — is principally a matter of reflection upon the ways in which language disguises our embodied way of thinking. But this is a matter of the explicit and formal character of what it is to follow a rule. It is a matter of showing us how the quandries and puzzles which have perplexed philosophers over the centuries — there is an interesting parallel to Heidegger here — dissolve when we come to grasp the self-grounding character of practice — in Wittgenstein's words how practice "takes care of itself" (*OC*, §139). This dissolution, it should be remembered, will be on the basis of an understanding of our natural history, which is to say, wholly compatible with the Naturalist demand that we come to see the evolutionary basis of knowledge and mind, but on the basis of "general facts of Nature", rather than a theory. Once we understand their significance, there is no longer a point to continue philosophizing: our perplexed meanderings simply come to a stop.

Here it is worth noting that the hermeneutics of the pargmatist metaphysician is very peculiar indeed. For him philosophy has to be a most peculiar activity. It is an investigation into questions which have no answers (Cf. *Tractatus*, 6.5). He takes traditional metaphysicians to be engaged in an activity that does and should resemble madness from the perspective of 'normal' people. (*OC*, §467). However, what is even more astonishing is that his own activity must be madder yet than theirs, if he is to solve their problems (*C&V*, p. 75). It is for this reason that our philosopher, unlike the Naturalist, and reminiscent of Zarathustra, can *never* be a member of a commnity of ideas (*Z*, §455).

Two final points need to be stressed here. First, there is, of course, no hope of 'refuting' the Naturalist in any of these matters; for it is curiously part of the common heritage of these two views of philosophizing that there is no such thing as an *experimentum* — or *argumentum* — *crucis*. So, these considerations are advanced as matter for reflection. Second, the case for critical hermeneutics that I have presented is clearly set forth here on the basis of metaphysical considerations; however, the value of critical hermeneutics can only be established pragmatically, which can only mean on the basis of successfully carried through critical hermeneutic *exercises*. The proof of the pie is in the eating — and we should never forget it. The most that I can hope to have accomplished here is to show the woeful incomprehension that attends J. B. Thompson's notion that, because Wittgenstein's later philosophy is committed to leaving everything as it is, it is by no means for all that uncritical; for we have a dogged, if wrong-headed, tendency not to be able to see just that.[19] But to have put a chink into the armor of a Habermasochist knight errant is very different from finding the Grail.

Let met close by returning briefly to Strindberg. In making the overdetermined nature of Miss Julie's behavior the focus of his play, Strindberg was certainly suggesting a certain perspectivism with respect to the interpretation of action:

> A suicide is committed. Business troubles, says the man of affairs. Unrequited love, says the woman. Sickness, says the invalid. Despair, says the down and out. But it is possible that the motive lay in all or none of these directions, or that the dead man concealed his actual motive by revealing quite another, likely to reflect more to his glory.[20]

Yet, it should be clear that Strindberg by no means seeks a single 'scientific', explanation which accounts for all of these. Rather, he introduces skepticism on the basis of overdetermination where his spectators expect to find themselves certain of what transpires. His aim is not to pander to, but to defeat, the uncritical attitide of the simple-minded. His use of overdetermination in aid of a perspicuous representation of motivation is very much an exercise in critical hermeneutics.

A NOTE ON *ABRICHTUNG*

That human education originates as a matter of *Abrichtung* or dressage should not by any means be construed as implying that it ends there. In the course of presenting different versions of the view presented here Stanley Cavell and Gerhard Frey have questioned respectively my associating coaching with drill sergeants and the matter of making the education process independent of the child's capabilities. I take these objections to be two aspects of the same point: eduction has to be more than a matter of regimentation.

Nothing in the view that I have propounded here conflicts with their humanistic concerns. Wittgenstein's point is that its ineliminable basis is our relation to nature, which includes both dressage and invention. Dressage starts as a reaction to pain. The child does not put its hand back into the fire again once having been burned. This sort of dressage is an ineliminable component of any further education given the kind of beast that we are. Further education is possible only because we have this 'instinctive' relation to nature. This relation has been eloquently articulated by Professor Malcolm.

The good coach is precisely the one who understands how to distinguish between those points in the development of novices when they most need drill and those points at which friendly encouragement is what is required. It is a matter of expertise that defies systematization because it is based upon an estimation of an individual's development. The point is that drill is by no means incompatible with the sort of "humanistic" attitude to education that Cavell and Frey allege to be lacking in my presentation of Wittgenstein. Drill too belongs to our "natural history", i.e., to growth and development. Indeed, it is only after having mastered a technique that we are capable of practicing the acquired skill inventively.

Abrichtung and invention are, then, two ends of a spectrum along which we pass as we develop a particular capacity. A competent coach or teacher is that person who mediates the experiences of a community in the best way. This is a matter of performance and certainty embraces what Cavell and Frey allege is missing in my presentation. The point is that the sort of *Abrichtung* Wittgenstein emphasizes is compatible with *all* forms of education. It is the indispensable basis upon which all further learning depends. It passes, 'dialectically', as it were, into guessing and then inventing ways of further developing the practice with the help of hints and tips from the coach.

NOTES

* I should like to acknowledge the stimulation I have received from conversations with Wolfgang Lackner and Bo Göranzon on topics relevant to this paper. They, like my other interlocutors, bear no responsibility for its shortcomings.

1 August Strindberg, Author's Forward, *Miss Julie*, trans., Elizabeth Sprigge (Garden City, 1955), p. 63.

2 The point of departure for my critique of Naturalism is the work of Owen Flanagan, in particular *The Science of the Mind* (Cambridge, Mass., 1984) and "Consciousness, Naturalism and Nagel" *Journal of Mind and Behavior*, Vol. **6**, no. 3 (Summer, 1985), 373—90. I am grateful to Owen Flanagan for innumerable discussions over the last ten years.

3 Flanagan, "Consciousness", 378—9.

4 See Flanagan, *The Science of the Mind*, pp. 98—104.

5 Stephen Toulmin once remarked in conversation that he found the whole concept of a psychological experiment very curious and questioned whether there could in fact be such a thing. I do not know whether he meant to say something along the lines I have suggested here. My point is that it is often (not always) unclear what we ought to infer on the basis of them. For Naturalists it is a foregone conclusion that they tell us something important because they have to be our *exclusive* avenue of access to the

mind. This is what I have set out to call into question. Ironically, it is in part the very brilliance of Flanagan's analyses that have led me in this direction.

[6] Flanagan, *The Science of the Mind*, pp. 66—74.

[7] Alasdair MacIntyre, *The Unconscious* (London, 1958).

[8] Flanagan, "Consciousness", 386 *et passim.*

[9] I owe this point to Philip Latey.

[10] Aristotle, *Nichomachean Ethics*, trans. W. D. Ross (Oxford, 1915), II, 2, 1103b; IX, 9, 1169a ff. I have drawn some of the implications of this view as it contrasts with what philosophers currently expect from ethics in my paper, "Does Ethics Rest Upon a Mistake?", presented to the Bergen University Philosophy Department in October, 1983 (unpublished). Flanagan concludes his "Admirable Immorality and Admirable Imperfection" *Journal of Philosophy* LXXXIII (1986), 41–60 on a similar note to my point of departure in a treatment otherwise wholly different.

[11] B. F. McGuinness, "Freud and Wittgenstein", *Wittgenstein and His Times*, B. F. McGuinness (Chicago, 1982), p. 27.

[12] Ludwig Wittgenstein, *Lectures and Conversations on Aesthetics, Psychology and Religious Belief*, ed. Cyril Barrett (Oxford, 1967), p. 42. Hereafter I refer to Wittgenstein's writings parenthetically in the body of my text as follows:

Tractatus = *Tractatus Logico-Philosophicus*, trans., D. F. Pears and B. F. McGuinness (London, 1961);

PI = *Philosophical Investigations*, trans., G. E. M. Anscombe (2nd. ed.; Oxford, 1969);

Z = *Zettel*, trans. ~~G.~~ E. M. Anscombe (Oxford, 1967);

OC = *On Certainty*, trans., G. E. M. Anscombe and Denis Paul (New York, 1972);

C&V = *Culture and Value*, trans., Peter Winch (2nd ed. Oxford, 1980).

[13] See Norman Malcom's penetrating essay "Wittgenstein: the relation of language to instinctive behavior", Philosophical Investigations, Vol. **5**, no. 1 (1982), 3—22. I have profited greatly from conversations about the relation between causes and reasons in the explanation of behavior in the later Wittgenstein with Kjell S. Johannessen.

[14] For the classical statement of this view see Rudolf Carnap, "Psychology in a Physical Language", *Logical Positivism*, ed., A. J. Ayer (New York, 1959), 165—98.

[15] On Darwin and evolution by natural selection see Loren Eisley, *Darwin's Century* (Garden City, 1961), 141—204.

[16] Aldous Huxley, *After Many a Summer* (Harmondsworth, 1955), 184—85.

[17] This is stressed, for example, in the works of John Ziman. Stephen Jay Gould has explored the darker side of this picture in *The Mismeasurement of Man* (New York and London, 1981).

[18] Nancy Mairs, *Plaintext* (Tuscon, 1984). Mairs's book is an autobiographical account of the way in which feminist identity can be, and in her case was, perceived as madness in patriarchal society. This is, curiously, the sort of thing Wittgenstein seems to have perceived as deeply problematic in modern thinking about mental disturbance: "madness *need not* be regarded as an illness. Why should'nt it be seen as a sudden — more or *less* sudden — change of character?" (*C&V*, p. 54).

[19] J. B. Thompson, *Critical Hermeneutics* (Cambridge, 1981), p. 34.

[20] Strindberg, *loc. cit.*

CHAPTER XII

PSYCHOANALYSIS: SCIENCE, LITERATURE OR ART?

"Es klingt wie ein wissenschaftliches Märchen," with these words on the evening of April 21, 1896 Count Richard von Krafft-Ebing dismissed Freud's account of the aetiology of hysteria as the result of childhood rape and thereby opened up vehement controversy with respect to Freud's work, which has persisted to this day and shows precious little sign of resolution.[1] In the many and varied controversies over psychoanalysis between then and now hardly an aspect of Freud's work has escaped becoming the focus of more bitter conflict. For the most part the reception of psychoanalysis in Austria has been exactly the opposite of its reception in the United States (and, more recently, in France). Austrians have tended to reject it out of hand; whereas Americans have embraced it wholeheartedly and unquestioningly. In both cases the general public's attitude has been less than critical: psychoanalysis has usually been extolled or excoriated almost always for the wrong reasons. In what follows I am less interested in concocting a catalogue of the Austrian friends and foes of Freud's movement than I am with suggesting — there can be no question in such a short piece of establishing — the actual relationship between its scientific promise, its therapeutic performance and its mythopoeic character on the basis of some of his Austrian critics. I shall indicate why I find Arthur Schnitzler's critique of psychoanalyisis especially noteworthy.

To be sure, Freud's claims that psychoanalysis was in fact the science of the mind aroused hostile reactions and wild enthusiams that were in direct proportion to one another. The dogmatism of the true believers among Freud's followers was matched by critics such as Karl Popper and Karl Kraus. For the former psychoanalysis was dangerous because it could not be tested; whereas, for the latter the idea that scientific investigation could penetrate the depths of human nature was abhorent. Both the view that the main tenets of psychoanalysis are so formulated as to be untestable and the view that it is the disease for which it claims to be the cure, while memorable and not entirely off the mark, are at once, overstated and vague: overstated because, as a matter of fact, no account of human behavior has been subject to more critical scrutiny than it has; vague, because poking fun at its excesses and pretensions, however justified, does not help us to assess its merits.

Ferdinand Ebner is a subtler, and generally more sympathetic, critic of psychoanalysis than either Popper or Kraus, while by no means opposed to the sorts of observations they bring to the subject. It is not necessary to accept the Christian standpoint Ebner endorses to appreciate his position with respect to psychoanalysis — although in fairness it must be said that neither is it easy to disentangle the two. Ebner distinguishes a profound element in the practice of psychoanalysis, insistence that the illusions of transcendence which pervade our waking lives (what he called the *Traum vom Geist*), whether moral, aesthetic or political are every bit as much dreams as what transpires during sleep, from that self-defeating element in psychoanalytical theory which is indeed the disease rather than the cure, the notion that psychoanalysis alone can "cure" us of that propensity. In the end Ebner rightly sees that the theory and practice of psychoanalysis are at odds with one another; it is capable of liberating us from one set of illusions, but no less capable of instilling in us an ideal of an alternative way of living which is no less unreal in its dream-like qualities.[2] Regardless of our judgment with respect to Ebner's religious commitments, there can be little doubt that he was among the first to appreciate both the strength and the weakness of psychoanalysis. It is the merit of Wittgenstein and Schnitzler that they press what is essentially the same sort of critique yet further.

For Wittgenstein Freud was one of the few authors who was genuinely worth reading despite the slavish attitude Freud inspired in his immediate circle. Moreover, he went far enough as to describe himself as a disciple of Freud's (interestingly he never described himself as being "influenced" by him as he did by others including Otto Weininger).[3] It was not that Wittgenstein agreed with Freud's views or theories; rather it was that he admired the assumptions Freud made in the practice of psychoanalysis, especially those concerning the way in which we are tempted to deceive ourselves in our efforts to attain self-understanding, Freud's emphasis upon the way that self-understanding only gradually dawns upon us as we come to introject a new scenario onto the trajectory of our behavior and, above all, the way Freud would have us resolve our problems by accepting a deflationary — and therefore only painfully attained — picture of ourselves. Wittgenstein was able to see clearly, as Ebner did only in an oblique way, that there was never any special reason why this process of self-interpretation should stop except arbitrarily — something he took to be an important criticism of psychoanalysis.

It is a great pity that Arthur Schnitzler never organized and articulated the notes he made over the period 1904—1926 into a full-fledged critique of psychoanalysis.[4] Of all of Freud's contemporaries Schnitzler was far and away the most qualified to pronounce upon the new would-be science. Not only was Schnitzler's medical background virtually identical with Freud's — his *Über funktionelle Aphonie und deren Behandlungen durch Hypnose und Suggestion* was a serious anticipation of the celebrated study of Freud and Breuer[5] — but his powers of observation and narration, his sensitivity both to human suffering and self-deception and his humanistic commitments were every bit the equal of Freud's. This being the case, his skepticism with respect to some of the central claims of psychoanalysis, for all its fragmentary character, is especially noteworthy.

To be sure, Schnitzler recognized that psychoanalysis was after something most important indeed. No less than Freud he was convinced that there was indeed a causality which governed and unified psychic life — in what seems to be an ironic reference to Mach, Schnitzler speaks of the economy within the human soul — but precisely for that reason he raised doubts concerning the adequacy of the psychoanalytic account of mental life. His criticisms of Freud may be classed under five rubrics. First, Schnitzler takes Freud to be precipitous in moving from the conscious to the unconscious but also with respect to his generalization of the Oedipus Complex. Second, he finds many psychoanalytical pronouncements simply sanctimonious reformulations of banalities such as the idea that our first sexual impressions are tied to our first relations with a person of the opposite sex. Third, he criticizes the inaccuracy of Freud's understanding of the Oedipus myth and its inappropriateness as a scenario or paradigm for the sort of father-mother-son relationship he wants to articulate. Fourth, Schnitzler considers Freud's work as oversimplifying crucial distinctions such as those between sexuality and eroticism, *Seelenkrankheit* and *Geisteskrankheit* as well as the notion of libido inasmuch as he restricts it to the experience of pleasure without allowing for the way in which inflicting pain can be a source thereof. Finally, Schnitzler sees pschoanalysis as shot through with arbitrariness with respect to the interpretation of dreams, folklore, myths, and pre-history but especially with respect to the way in which it projects metaphors such as *das Es, das Ich* and *das Über-Ich* in the course of mapping the psychic terrain, that are in turn discussed as though they were clearly and distinctly identifiable components of the psyche. Schnitzler has no doubt that this distinction is ingenious but nevertheless artificial. Thus, we arrive at the core of his objection to psychoanalysis: Freud casts a "shape" on the mind, which seems to Schnitzler to distort the fluidity of the relationship which in fact obtains between the conscious and the unconscious.

In contrast to Freud, and with a higher degree of phenomenological accuracy on his own view, Schnitzler insisted that the psyche would be better understood as composed of conscious (*Bewußtsein*) and subconscious (*Unterbewußtsein*) but only as separated by a vast preconscious (*Mittelbewußtsein*). In a short passage, which is as frustrating as it is tantalizing, he compared the difference between the two to the difference between entrancement in the literal sense and sleep. It is easy to wake someone normally from the latter; whereas the former is more complicated. Briefly, Freud's concept of mind was not dynamic enough for Schnitzler. The latter emphasized the way that things flit into and out of conscious awareness (the similarities with William James's notion of the stream of thought with its ever changing eddies and fluctuating relationship between periphery and center press themselves upon the informed reader — Schnitzler could well have been aware of James's work, which was very well received in Central Europe at the turn of the century). It does not seem unjust to describe Schnitzler's plays as explorations of this pre-conscious on the basis of the way in which they emphasize how we choose to be aware of — or to discuss — certain parts of experience at the expense of others, on the one hand, and, on the other, how those parts of experience we deliberately ignore have a way of pressing themselves upon us in unexpected and traumatic ways. I think it is worth pointing out that Freud's early claims on behalf of the merits of self-analysis as a mode of confirming his views would seem to be highly compatible with Schnitzler's notion of a pre-conscious to which we have access but only with difficulty. Be that as it may, Schnitzler's less metaphysical, more supple account of psychodynamics finds interesting reverberation and confirmation in two very different scientific critiques of Freud: those of Alasdair MacIntyre and, more recently, those of Jeffrey Masson.

MacIntyre's work represents an explicitly Popperian approach to psychoanalysis — not the crude falsificationism of Popper himself but the sophisticated methodological approach which Popper should have taken to the subject.[6] The intracasies of MacIntyre's argument are such that they defy summary. What is important here, however, is that MacIntyre makes a compelling case for separating the truth claims of psychoanalytic theory from the successes of psychotherapy. For the therapy to work the theory need not be true; it is only necessary that the patient believes that it is true — something which distinguishes the practice of psychotherapy from the practice of medicine, whose successes depend upon the skill of the doctor rather than the patient's belief system. In fact, MacIntyre claims, the successes of psychoanalysis are tied to convincing — or better converting — the patient to accept Oedipal motivation as the ultimate explanation for those actions which have become enigmatic to him. On this account, the distinguishing characteristic of psychoanalysis is its mythological dimension, i.e., the role that story-telling plays both in its theory and therapy; for no small part of Freud's achievement, when his work is so construed, consists in his literary accomplishment. We ought to remind ourselves that *Die Traumdeutung* won him the Goethe Prize for Literature in 1900. One need only compare any of his case histories with those of his predecessors such as Krafft-Ebing to see that his diagnoses involve a fuller and richer narrative by far than theirs. We forget at our peril that much of Freud's ability to convince at the theoretical level rests upon the fulness and richness of the hermeneutic which he made available as a scenario for self-understanding for people who desperately required just that. Indeed, the literary character of his work was something that seemed to bother Freud. In a moment of self-doubt in 1895 ". . . it still strikes me as strange that the case histories I write should read like short stories".[7] Moreover, the success of Freud's therapy to a great extent depends upon the patient's ability to tell himself a new story about his life in such a way that the Oedipal situation comes to illuminate the spiritual impasse which drove the patient into therapy in the first place (the importance of this sort of mythopoeic introjection has been emphasized by Roland Barthes). By effectively sundering theory from therapy and recasting therapy as a form of myth-making MacIntyre's conceptual analysis strongly supports the views of Ebner, Wittgenstein and Schnitzler. Recent revisionist accounts of how Freud came to espouse the metaphysics of the Oedipus Complex tend to strengthen that view yet more.

In a highly controversial book researched during his tenure as keeper of the Freud Archives (a job that publication of his thesis cost him), and unfortunately for anyone with healthy Popperian instincts, on the basis of archival material unavailable to the public in the lifetime of any currently living human being, Jeffrey Masson has claimed that Freud developed the Oedipus Complex as an alternative to the seduction theory of the aetiology of hystrical traumata, less on the basis of facts and convictions, than on that of banal careerism.[8] On Masson's view Freud was convinced on the basis of what happened on that fateful night in April 1896 that his medical colleagues would never take seriously the idea that children were actually raped by their parents. Thus, in place of the actual event he convinced himself — in good faith, it seems — that what was reported as fact ought to be construed as fantasy. If Masson is right, Freud talked himself into accepting the view that patient's stories about childhood rapes were fanciful and erected the metaphyscial edifice of the Oedipus Complex to explain their fantasies precisely because he thought that this was the only way that he could ever be accepted by the medical profession. Thus, we have at least reason for suspecting that just that element which has provoked the most penetrating critiques of psychoanalysis was, in effect, a mistake from the start. If Masson is right, Freud substituted a yet more far-fetched, but more socially acceptable scientific fairy tale for the one that Krafft-Ebing and his colleagues so silently and unequivocally rejected in 1896. My intention here is neither to defend Masson's thesis nor to speculate upon whether psychoanalysis could have been developed without the notion that childhood rapes were wishes and not real events, but the very idea that there was a plausible alternative to the scenario which has been the chief point of attack by the most sophisticated critics of psychoanalysis, the idea of an alternative which involved neither the metaphysics of the Oedipus Complex nor the sharp distinction which Schnitzler objected to between the conscious and the unconscious, seems generally to support the sorts of criticisms of Freud discussed here. If there is anything at all to Masson's thesis, it certainly provides aid and confort for those critics of psychoanalysis for whom Freud's alleged science represents a more or less successful therapeutic myth — it must be added that the obverse is also true, i.e., that the kinds of conceptual point which Ebner, Wittgenstein and Schnitzler want to make about psychoanalysis tend to support, not the facts which Masson rests his case upon, but the *plausibility* of his general position.

What we know about psychoanalysis with a modicum of certainty, then, seems to bear out the views of several of Freud's Austrian critics. Maybe psychoanalysis is a science, but in truth we hardly know what a science of the mind with the kind of explanatory power that Freud would demand would look like. Psychoanalysis is certainly a healing art which is at least as good as any alternative.[9] However, that very therapeutic success rests upon a mythopoeic element, which, in the end, may be its most important component. In at least one respect there is universal agreement with Schnitzler: Psychoanalyse ist scheinbar eine leichtere, in Wirklichkeit eine schwerere Kunst als man glaubt".[10]

NOTES

[1] Quoted by Jeffrey Mousaieff Masson, *The Assault Upon Truth* (New York, 1984), p. 9.

[2] Ferdinand Ebner, *Das Wort und die quistige Realitäten* (Frankfurt/Main, 1980), p. 262.

[3] Ludwig Wittgenstein, "Gespräche über Freud", *Vorlesungen und Gespräche über Ästhetik, Psychologie und Religion*, hrsg. Cyril Barrett; übers. Eberhard bubser (Göttingen, 1968), pp. 73—86. Brian McGuinness, "Freud and Wittgenstein", *Wittgenstein and His Times*, ed. B. F. McGuinness (Chicago, 1982), pp. 27—43.

[4] Arthur Schnitzler, "Über Psychoanalyse", *Protokolle* (1976), pp. 277—84.

[5] Bernd Urban "Arthur Schnitzler und Sigmund Freud: Aus den Anfängen des "Doppelgängers", *GRM* XXIV (1974), pp. 193—223.

[6] Alasdair MacIntyre, *The Unconscious* (London & New York, 1958).

[7] Quoted by MacIntyre, p. 43.

[8] See note #1. For an account of the controversy surrounding Masson's work see Janet Malcolm, *In the Freud Archives* (New York, 1984).

[9] This is argued persuasively by Seymour Fisher and Roger P. Greenberg, *The Scientific Credibilty of Freud's Theories and Therapy* (New York, 1977).

[10] Schnitzler, *op. cit.* p. 280.

CHAPTER XIII

BETWEEN ENLIGHTENMENT AND COUNTER-ENLIGHTENMENT: THE SELF-CRITICAL RATIONALISM OF G. C. LICHTENBERG

Why are young widows in mourning so beautiful? (Investigation)[1]

The question amuses and charms in its naughty way. But *are* young widows in mourning *so* beautiful? Perhaps. The question more than insinuates that they are; while the parenthesis entices you to dwell upon the question. And it is precisely this insinuating, enticing character that is crucial here; for the epigram not only congers up an image of sweet young defenselessness, but also teases you into reflecting upon that image. But at this point something special happens. If we really consider the issue, we discover that the putative beauty of young widows in mourning resides principally in our imagination, indeed, it is our sexual fantasy that congers up youthful femininity made beautiful by its vulnerability. What starts out as a witticism with a bit of concentration turns out to be nothing less than a way of suggestively exploring our pre-conscious psyche. What is on the face of it a clever slightly suggestive aperçu through its very innuendo suddenly transforms itself at a deeper level into a (potentially embarrassing) insight into our own lust. In the end we attain a surprizing, slightly deflating insight into ourselves, whose power to deflate rests precisely on its exploitation of our very power and willingness to fantasize. The aphorism combines consummate insight with as much artistry. It is typical Lichtenberg.

Georg Christoph Lichtenberg (1742–1799),[2] like the Diderot of *Rameau's Nephew*,[3] represents a crucial phase in the development of the Enlightenment: that point — equidistant between Voltaire and Rousseau — at which Enlightenment comes to realize its inherent limitations without yet having become Romanticism. Lichtenberg, apart from being an aphoristic moralist of the first order, was a first-rate mathematician, astronomer and experimental physicist by training. However, like the author of *Rameau's Nephew*, he became deeply skeptical with respect to the Enlightenment's hyper-rationalistic tendencies. Like Diderot, Lichtenberg by no means abandoned reason for faith, as did Hamann, or feeling, as did Rousseau and the Romantics. He, rather, strove to clarify the limits of Enlightenment. His question was 'what can we expect from the exercise of reason?'. At the outset of his career Lichtenberg was a convinced Leibnizian rationalist scientist. Educated in Göttingen at the finest German university of the day, he was thoroughly imbued with the spirit of Leibniz' efforts to establish the a-priori character of all of knowledge including physical science (he never lost his fascination with the idea of an absolutely precise universal language, despite rejecting it), Lichtenberg's practical, experimental cast of mind soon made him into a modest skeptic with respect to all such visionary projects without turning him against the sorts of *question* such speculations would answer. In particular, Lichtenberg became fascinated with the process of scientific *discovery*.[4] It was the very impossibility of predicting discoveries that turned him away from rationalist system-building. If nature could be grasped in an a-priori system of concepts, radical conceptual change, genuine novelty, would not exist. But the *developmental* character of science wholly contradicted this rationalist assumption. No algorithm for empirical science was to be found. His problem became one of accounting for the way in which *wholly unforseen* developments in science could take place. It is important to emphasize that, while Lichtenberg became dissillusioned with all forms of *scientism*, i.e. with all efforts to confer normative value upon science with respect to human activities, at a very early stage in his development; he remained a practicing scientist throughout his life. Moreover, having become one of Europe's most distinguished teachers of science, he did his best work in the philosophy of science late in life. His philosophy of science was wholly rooted in coming to understand the nature of scientific *practices* — a perspective foreshadowing William James and Thomas Kuhn: "In order to see something new, you have to make something new".[5]

Lichtenberg's view of science, then, is strikingly modern in at least two respects. First, he sought to understand scientific practices in terms of the way they are constituted through the scientist's *language*, rather than as mathematical system or an assemblage of 'sense data'. Radical conceptual change, the emergence of the unforseen, was tied to the deployment of novel *analogies or metaphors*, which produced new ways of seeing natural phenomena, which, in turn, allowed the scientist to organize his research more precisely than had hitherto been possible. The second strikingly contemporary note about this is that he described those practical models upon which scientific reasoning in the concrete rested as *paradigms*. They were the key to wresting her secrets from a reluctant nature. By inventing new analogies we, in effect, produce a new grammar of investigation as we seek to "decline" our paradigm. However, it is precisely the implications of this concept of science for 'enlightenment' which make his views, not merely startlingly relevant to our debates today, but absolutely crucial to them.

First of all, it led Lichtenberg to a methodological pluralism (without undermining his confidence in the unity of nature) of a sort that has only become respectable on a large scale in the last quarter century. Thus he could refer to "the myth of the physicists".[6] Secondly, it made him conscious of the importance of the individual, as opposed to the general, and the small, as opposed to the large: "the real trick is to see the significant in the small. If sagacity is a magnifying glass, wit is a minifying glass. Do you believe discoveries can only be made with a magnifying glass? I believe more discoveries have been made in the intellectual world with a minifying glass or at least with similar instruments".[7] This was certainly the method that Lichtenberg himself employed. The ways in which it anticipates a similar strategy, that of breaking language up into small cross sections for the sake of illuminating nuances easily passed over in normal discourse, the method of distinguishing various "language games" which Ludwig Wittgenstein employed in his *Philosophical Investigations*,[8] should not pass unnoticed. They are hardly accidental. Thirdly, Lichtenberg came to see scientific knowledge more as a matter of *invention* than discovery — or, to put the matter slightly different, he came to see the observed and the observer as inextricably bound together. Inasmuch as knowledge was rooted in analogy and metaphor, the creative scientist was more of a poet than a logician formalizing the 'merely' empirical or a bureaucrat certifiying the empirical validity of claims about the world. It is important to emphasize that Lichtenberg saw this more as a way of cajoling, persuading (seducing?) nature to give up her secrets than of 'inspired' guessing: "the great knack of regarding little deviations from the truth for the truth, upon which the whole of the differential calculus is built, is also the basis of our witty mode of thought, where the whole thing would fall to pieces, if we were to regard the deviation with philosophical rigor".[9] Thus, Lichtenberg would not have been terribly surprized or upset to discover that, say, Mendel's ratios were a bit 'too perfect'. Be that as it may, if that is an adequate description of science as it is practiced, there could be no question of conceiving reality as a 'rational' system. For one thing, if knowledge is a construct, there is a sense in that nature can. This implies an end to rationalistic philosophizing, but no less poses a difficult problem for our understanding of the self or the philosophy of mind. Our knowledge of mind cannot hope to have the same form as our knowledge of nature. Precisely because our knowledge of nature is dependent upon the conceptual networks we create, we cannot hope to catch that phenomenon in those nets. It is important to emphasize that this is not to say that there is no experimental or scientific way of getting at the workings of the mind for Lichtenberg (at least as he would understand those terms), but that the self-reflexive character of those methods will be radically different from those of the physical sciences. The methods must be odd, because the task is odd. It involves nothing less than getting a glimpse of ourselves as we construct our modes of seeing. The trick is roughly the equivalent of playing the piano and watching oneself do so in a mirror — a very special talent indeed. For this very special task an equally special medium was required. The aphorism was his means of giving himself — and us — the requisite second look at ourselves. It will be necessary to say a lot more about this.

Finally, one of the most important consequences of Lichtenberg's confrontation with the problem of scientific method follows from his reflection on the way science is rooted in language. If the scientist's primary tool is the analogy the distance between scientific method and other human activities narrows considerably. If the core of scientific method is analogy, science is continuous with what in English is praised as common sense or ordinary practical reasoning. The latter is not merely a provisional starting point or a genetic element to be eliminated at some later, more refined stage of development, but the most fundamental determinant of scientific method. 'Epistemological breaks', on this account, turn out to be illusions rooted in confusing scientific reasoning with its mode of presenting its results — Lichtenberg, like Stephen Toulmin, was convinced that the most important stage in the development of a physical theory took place *before* we expressed ourselves in the form of equations. But, if all of this is right, science cannot be the measure of rationality. Contrary to the views, say, of W. V. O. Quine, our scientific practices are distinguished from other sorts of rational procedures on the basis of their precision and organization, as well as by virtue of the fact that the community in which they are developed shares criteria for and commitment to debating theses to a far larger extent than society at large, but those practices are in no way qualitatively different from the social practices in which they nest: "The philosophy of the common man is the mother of ours; our religion develops from his superstition; just as our medicine from his healing craft".[10] This view, however, entails a revision of what we should expect from science. The main stream of the Enlightenment and the French Revolution notwithstanding, we should not expect it to transform society; for it is itself rooted in a certain type of social values which Lichtenberg took to be more widely cultivated in England than elsewhere. Science can only transform society to the extent that said transformation is compatible with traditional values. The latter incorporate a rationality of their own, which, Lichtenberg recognizes, is not identical with the claims that traditionalists make on behalf of tradition.

Lichtenberg's discovery of the basis of scientific practice in analogy had the effect of de-mythologizing science in his eyes. However, this should by no means be taken to signify that he became any less interested in science on that account. Rather, it had the effect of turning him into a critic of the Enlightenment's tendency to idolize science as well as the pretentions of certain scientists. His confrontation with the pretentions of science centered principally upon his polemics against Lavater's physiognomy. Lavater believed that character was expressed in the shape of a person's head. Thus, from a silhouette he claimed to be able to analyse personality traits. This was an interesting problem for a committed empiricist like Lichtenberg precisely because certain claims about a person's psychic constitution were being made on the basis of putatively empirical information about that person. Today we would say that the problem Lichtenberg wanted to draw attention to was a conceptual one. It was not that the physiognomist was not empirical enough; rather, that he chose irrelevant data in his efforts to analyse character: "it is peculiar, and I have never noticed without laughing, that Lavater finds more on the nose of current young writers than reasonable people do in their writings".[11] In fact, for all Lavater's empirical data, he had developed a crude form of metaphysical determinism and paraded it as a new empirical key to psychology (we forget at our peril that Enlightenment commitment to science at a point in history when most disciplines were in a relatively primitive stage of development often was an an encouragement to wild speculations as in the case, say, of Mesmer). For the self-critical rationalist, Lichtenberg, it was crucial to show that not everything calling itself science was worthy of that name. So, he turned his skeptical wit upon Lavater & Co.: "Chiefly people with completely dissimilar faces are often otherwise very similar. At least the relation of their temperaments is not in their faces".[12] Here, as elsewhere, Lichtenberg placed great faith in traditional beliefs as foils to pseudo-scientific fads. It is of the essence of Lichtenberg's self-critical rationalism that he opposes any simple-minded, mechanistic effort to lay bare the workings of the mind. His aim was nothing less than to create a genuinely experimental study of the mind. For Lichtenberg this had to be a very peculiar undertaking, since the subject of study included the researcher himself, as well as the language he must use in his investigation. Thus, the aphoristic thought experiments which constitute his oeuvre are a peculiar, and peculiarly profound, critique of language and an experimental logic, as well as a major contribution to an experimental science of the mind.

As we have seen Lichtenberg's understanding of natural science was rooted in an understanding of what it was to experiment. In wresting knowledge from a recalcitrant nature, "the world in not there in order for us to know it",[13] language was the key. However, nothing is more characteristic of his enlightened sense of the limits of Enlightenment that his double awareness of both our dependence upon language *and* the tendency of language to confuse us: "our false philosophy is embedded in the whole of language. We cannot reason without reasoning falsely as it were".[14] For this reason 'experimental' thinking must be a critique of language. However, it is important to stress that Lichtenberg understands the way in which language conceals reality from us along the pragmatist lines of the later Wittgenstein, rather than along the positivist lines of Ernst Mach (both of whom owed a great deal to him). When he wrote, "who is there? Only I. Oh, that's superfluous enough",[15] he protested less against the mere 'hypostatization' of a noun than he did against the ways in which the notion of an ego is a distorting mirror through which we are bound to misconceive what we really are. The ego of the philosophers directs reflection in just the wrong way, i.e., away from those experiences which we typically identify ourselves with most readily. If it was a mistake to conceive a self in the way that rationalist metaphysics and traditional theology did — and Lichtenberg certainly thought that it was — just what is it that tempts us in this direction? His task as an experimental psychologist was to explore the sources of our temptations in language and experience. Given his view of the constitutive role of language this was in fact one task. In effect his aphorisms became a sort of self-test in both the sense of a self administered test and a test of selfhood (Kierkegaard, aware of the fact that the Hebrew word for experiment was the same as the word for temptation was later to conduct analogous 'self-experiments'). This attitude to language made him skeptical with respect to all efforts explanatory theories in psychology. Because they accepted a phenomenology of action in ordinary language they were bound to go astray in their efforts to provide an explanation of action. For Lichtenberg, as for Wittgenstein much later, everything hung upon getting the description right in the face of temptations to go astray. Nobody was more tempted to do so than philosophers (there would have been no distinction between philosopher and psychologist in his day). Their mistakes consisted precisely in trying to tidy up our psychic vocabulary à la Mach or Skinner, without looking at the sorts of actions with which it is co-ordinated. "Everybody who speaks German is a folkphilosopher. Our university philosophy consists in making restrictions upon ordinary language. Our whole philosophy is correction of the use of language and thus the correction of a folk philosophy of the most general sort. The common philosophy alone has the advantage that is possesses declensions and conjugations. We always teach the true philosophy in the language of the false. To explain words doesn't help; for in explaining words I don't change pronouns and their declensions at all".[16] Thus emerges Lichtenberg's twofold task of polemicizing against theory in psychology and articulating what we understand of "ourselves" *in practice*.

Thus, the sort of psychology that Lichtenberg advocates is a refined sort of folk psychology in that he takes traditional *practices* as the embodiment of rationality. However, it is more than that inasmuch as he rejects the idea that what we normally say about our behavior should carry any special weight. In this spirit he could write: "man is a masterpiece of creation, among other things, by virtue of the fact that he believes, in spite of all determinism, that he acts freely".[17] For Lichtenberg *both* the determinism *and* our sense of acting freely were important data for psychology. He would not, with metaphysicians, both Rationalist and Empiricist, deny the latter on behalf of the former; nor would be naively reject science and metaphysics on the basis of popular belief. His aphoristic experiments were designed to test both science and traditional belief. He had only the greatest respect for both. But they needed each other. Traditions were the repository of experience. For that reason he looked askance at efforts to jettison tradition: "experimental politics: the French Revolution".[18] In the same vein he observed, "I fear our all-too meticulous breeding practices shall yield dwarf fruit".[19] His scientific task was to tease the rational core out of tradition.

Two things are crucial here for appreciating his method and the particular importance of style in this undertaking: the practice-immanent character of rationality and the distorting power of language. To explicate the former in the face of the latter he developed the aphorism as a form of thought experiment. Lichtenberg's thought experiments must be distinguished from those of, say, contemporary philosophers of mind, who proceed from the construction of certain hypothetical situations to examine their 'logical' implications. Lichtenberg's thought experiments are entirely different in character. They aim at laying bare our assumptions, that element in the grammar of our thinking which binds thought and action together. His strategy and tactics in this respect closely resembles those of Wittgenstein in as much as they constitute a technique for revealing to us precisely what is obscured to us by virtue of its familiarity. More in the spirit of Wittgenstein than, say, of Popper, he could write: "what is very odd remains seldom long without explanation. The inexplicable is usually what is no longer odd and, perhaps, never was so".[20] Lichtenberg found himself bitterly opposed to contemporary journalistic practices to the extent that they tended to treat everything they touched with just the sort of casual familiarity which obscured the profound in the everyday. In a similar Kierkegaardian vein he inveighed against fickle public opinion as the correlative of an irresponsible press. It was precisely the function of humor in his works to lay bare the the sublime element in the everyday, which journalism tends systematically to obscure: "wit is the discoverer; understanding is the observer".[21] Wit is a way of reminding us of the nuances that we have passed over too hurriedly in our efforts to 'explain'. An "Amen face",[22] a "slug of rationality",[23] "the real philosophers and the titular",[24] "a magnet, which falls in love with six pounds",[25] "the animal which drowns in a tear",[26] are all striking reminders to slow down and take a second look at what we take for granted in the world around us; but they are no less reminders of the way in which our very turns of phrase shape what we see when we look at anything. This double reflection, i.e. the capacity to produce multiple speech acts in a single turn of phrase, in Lichtenberg's work is hardly the least of his achievements. It too is strikingly contemporary.

Indeed, it would not be altogether wrong to consider him as a great-grandfather of a very significant stream within modern thought. This would be true even if his influence did not extend beyond Ludwig Wittgenstein and Friedrich Nietzsche to figures like Ernst Mach, who was indebted to him for his notion of the Economy of Thought and the unrescuable Ego, Karl Kraus and Ferdinand Ebner — the impact of Lichtenberg's thinking in Vienna at the turn of the century has yet to be studied in depth both with respect to its scope and the problem of explaining why he was particularly significant for Austrians when he was all-but unknown in Germany. It does not seem implausible that the answers to these questions will have a lot to do with what I have termed his self-critical rationalism. It will be particularly worthwhile here to examine Lichtenberg's relationship to the thought of Wittgenstein and Nietzsche to obtain a perspective on his historical importance as well as completing our picture of his peculiar non-Romantic mode of counter-Enlightenment.

The later Wittgenstein's works were distinguished in his own eyes, less for any originality with respect to their substance, than for the method they introduced for treating philosophical problems therapeutically. He acknowledged that this perspective on traditional philosophical problems was something that he adopted from psychoanalysis.[27] Doubtless, his concept of the *strategy* of philosophy unmasking false consciousness and convincing us to accept a less flattering picture of ourselves owed a great deal to other sources as different as St. Augustine and Karl Kraus, William James and Otto Weininger as well.[28] However, the roots of his philosophical *tactics* is, while acknowledged in the literature, less well-known and needful of further documentation. However, this in itself is not ground for avoiding the topic.

For one thing, Wittgenstein's mode of composing his works bears a striking resemblance to the method favored by Lichtenberg. The latter was of the opinion that we should all write down everything that catches our fancy. In this we should follow the lead of the merchant who scribbles down every transaction hurriedly on scraps of paper to be in a position later to enter them systematically into his ledger for the sake of precise reckoning of his accounts.[29] Thus, Lichtenberg's "waste books" were intended to be collections of raw material for novels (which in fact were never written). So, Wittgenstein recorded his pensées regularly on slips of paper, which he would later arrange and rearrange meticulously in the hope of finding the most natural 'logical' order for them. This is such an odd method that its similarity with Lichtenberg's preferred procedure is noteworthy.

More important, however, is his way of moving us to reflect upon what we have hitherto taken for granted about language and the world. Wittgenstein's recipe for making us take a second look at our assumptions, and, indeed, ourselves involved typically Lichtenbergian tactics such as the unanswered question (674 in the *Investigations*), striking aphorisms (a cloud of metaphysics, condensed into a drop of grammar) and, above all, thought experiments. They should lead us to a "perspicuous representation"[30] of our practices, one which our very involvement in them tends to obscure. This perspicuous representation is, on Wittgenstein's own account, a way of replacing explanation by description. However, the kind of description Wittgenstein aims at is hardly a mere recording of 'facts' about our practices. More often than not, it entails a 'perspicuous contrast' between our customary practices and imaginary ones, which contrast should clarify how our practices are rooted in our natural history. In order to make our natural history clear to us he had to invent silly-sounding fictional ones, which we could then compare with our own.[31] Thus, he would ask us why we talk about brown instead of redish-green as a heuristic device for bringing us to understand the order in our color-related practices in such a way that we would cease to pose 'philosophical' problems about them. It must be emphasized that the extent of Wittgenstein's debts to Lichtenberg is a matter which has yet to be systematically researched; nevertheless, the parallels to his thought experiments, only perhaps paralleled in Kierkegaard among the thinkers Wittgenstein is known to have admired, should not be neglected — not least because they mutually illuminate both thinkers. Think only of Lichtenberg's remark: "if there were only turnips and potatoes in the world, then one would perhaps say: its a shame that plants grow upside down".[32] The peculiar sort of invention (poetry?) we find in Wittgenstein's efforts to think crazier than traditional philosophers is certainly anticipated, if not actually influenced, by Lichtenberg.[33] We forget at our peril that the author of the *Philosophical Investigations* considered it of the utmost importance that his fanciful forays into fictionalized natural history — consider: 'the baby crawled for the first time today and the baby hoped for the first time today' — strike us as funny.[34] In this he was certainly one with Lichtenberg. But Lichtenberg's role as a precursor of Nietzsche is no less noteworthy.

Lichtenberg was one of the few German thinkers with whom Nietzsche could wholly identify.[35] Certainly, Lichtenberg's innumerable barbs at German crudeness, and his pellucid style, were mighty impressive to Nietzsche, despite the fact that England played the role in the former's thought that France did in forming the latter's cultural ideals (not to be confused with Ideals). Indeed, despite certain very significant differences in the orientation of their thinking, there is every reason for considering Lichtenberg among Nietzsche's most important precursors. Indeed, only someone ignorant of Lichtenberg would venture to claim that Nietzsche's style was absolutely unique. Moreover, we find in Lichtenberg clear anticipations of Nietzsche's critique of Winckelmann[36] of Eternal Recurrence (i.e., cultivating an outlook on life in which neither hope nor nostalgia played a major role)[37] the over-man,[38] the idea that language tends to trick us and, above all, the famous distinction between the Apollonian and the Dionysian components of life. A brief exploration of some of these similarities will help further to clarify Lichtenberg's position at the juncture of Enlightenment and Counter-Enlightenment.

True to the spirit of Enlightenment Lichtenberg sought to understand the world in causal terms. By now it should be clear that chief among the things he wanted to understand was his own psyche. Lichtenberg was, after all, a contemporary of Rousseau — and very consciously so. However, the sort of spontaneous selfhood Rousseau advocated was not so easily to attain. As we have seen, language laid traps for the would-be enlightened *Selbstkenner*; but no less our own desire to see ourselves other than as we are actually seen by others: "Every man has a moral backside (*sic*), which he doesn't show without necessity, and, which he covers as long as possible with the pants of good behavior".[39] Chief among our sources of self-deception is our sexuality: "what they call the 'heart' lies way below the fourth vest button".[40] In this Rousseauian effort to discover just who he was, Lichtenberg discovered — and affirmed — what Nietzsche would later term a Dionysian element in himself: "if nature had not wanted the head to give in to the demands of the belly, what would have been the need for putting a head on the belly in the first place?".[41] To understand how the whole man could move together (Lichtenberg's avowed goal as an anti-faculty psychologist) it is necessary to understand our place in nature. Sexuality is at once the most obvious link to nature and the one we are most prone to overlook and/or deny (i.e., before Nietzsche and Freud). As nature (despite Lichtenberg's admiration for Spinoza) is not there to be known by us but must be coaxed into divulging her secrets; so self-knowledge is something we can only wrest from nature with difficulty: "we, the tail of the world, don't know what the head is up to".[42]

Like Nietzsche — and Freud after him[43] — Lichtenberg was convinced that self-knowledge was tied to understanding our dreams, which he took to be continuous with our wakeful states: "I recommend dreams again. We live and experience just as well in dreams as we do while awake. And the former experiences are just as good as the latter. To dream and to know that you are dreaming is one of the advantages of being human. The dream is a life that is compounded with our other one, the one we call 'human life'. Dreams slip gradually into our wakeful states in such a way that we can't say where wakefulness begins".[44] To know ourselves is to come to know ourselves as dreamers. Therefore, Lichtenberg, like Bachelard, urges us to cultivate our capacity for dreaming. It, too, is an experimental method for exploring our mental terrain. Lichtenberg's presupposition is that there is more going on in our minds than we ourselves are aware of or want to be aware of; it is left to him, the aphoristic explorer of the mind, to lead us, and *himself* — for Lichtenberg in the last analysis there is no difference, i.e. whether one writes or reads the aphorisms — along the path of self-realization, just as he does when he/we ask why young widows in mourning are so beautiful. Lichtenberg's version of the Grand Style often entails leading us into what appears to be an Apollonian direction, only to discover that we in fact moved in quite another (Dionysian) one. "That man is the noblest of creatures", he writes, "may be inferred from the fact that no other creature has yet denied him this title".[45] On the surface a clear statement of human superiority. But reflect a bit and it strikes you that no other creature, as of this writing, is capable of telling him anything different. It is we who have made our sort of speech the criterion of our superiority. The aphorisitic form, by inviting reflection, expresses the vacuity of the reasoning involved in a movement of thought which makes its actual point a matter of realizing something that lies beneath the surface, something which requires a second look to be properly understood.

To be sure, it would be just as absurd to equate Lichtenberg with Nietzsche as it would with Wittgenstein however much he may have 'influenced' both. Nevertheless, it would be no less absurd to pass over the similarities between them (which help *inter alia* to explain similarities between Wittgenstein and Nietzsche!). No small part of the point of these comparisons is to underscore the relevance of Lichtenberg's way of thinking and writing for contemporary thought. His effort to find a happy medium between reason and feeling, progress and tradition, Enlightenment and Romanticism, as well as his awareness of the constitutive (transcendental?) role of language in our understanding of the world and ourselves, his emphasis upon analogy as the central feature of experimental thinking and his mode of investigating logical and psychological issues by means of thought experiments all address the most crucial issues in contemporary thought in a manner at once supremely humorous and humane, practical and profound.

NOTES

1 Georg Christoph Lichtenberg, *Aphorismen*, ed. Kurt Batt (Frankfurt/Main, 1976), p. 122. All references to the aphorisms are to this edition. Unless otherwise indicated. Unless otherwise indicated all translations are my own.

2 For biographies of Lichtenberg see the magisterial study of Franz Mauthner, *Lichtenberg: Geschichte seines Geistes* (Berlin, 1968), J. P. Stern, *Lichtenberg*: *A Doctrine of Scattered Occasions* (Bloomington, 1959), Wolfgang Promies, *Georg Christoph Lichtenberg* (Reinbeck bei Hamburg, 1964). I am particularly indebted to Stern in my presentation of Lichtenberg.

3 On *Rameau's Nephew* see Herbert Josephs, *Diderot's Dialogue of Language and Gesture*: *Le neveu de Rameau* (Columbus, 1969).

4 My presentation of Lichtenberg's views on science draws heavily upon G. H. von Wright, "Lichtenberg, Georg Christoph," *Encyclopedia of Philosophy*, ed. Paul Edwards et al. (New York, 1967), IV, pp. 461—65 and Stern's discussion of the scientific aphorism. Cf. Mauthner, *Lichtenberg*, pp. 365—85 *et passim*; Promies *Lichtenberg*, pp. 56—100. Lichtenberg gave his name to the electromagnetic phenomenon known as Lichtenberg "stars" or figures.

5 Lichtenberg, *op. cit.*, p. 204.

6 *Ibid.*, p. 178.

7 *Ibid.*, p. 78.

8 Ludwig Wittgenstein, *The Blue and Brown Books*, ed. Rush Rhees (New York, 1958), p. 17.

9 Lichtenberg, *op. cit.*, p. 9.

10 *Ibid.*, p. 59.

11 *Ibid.*, p. 132.

12 *Ibid.*, p. 138.

[13] *Ibid.*, p. 193.
[14] *Ibid.*, p. 152.
[15] *Ibid.*, p. 38.
[16] *Ibid.*, p. 152.
[17] *Ibid.*, p. 203.
[18] *Ibid.*, p. 237.
[19] *Loc. cit.*
[20] *Ibid.*, p. 210.
[21] *Ibid.*, p. 204.
[22] *Ibid.*, p. 135.
[23] *Ibid.*, p. 97.
[24] *Ibid.*, p. 107.
[25] *Ibid.*, p. 129.
[26] *Ibid.*, p. 89.
[27] Brian McGuinness, "Freud and Wittgenstein," *Wittgenstein and His Times*, ed B. F. McGuinness (Chicago, 1982), pp. 27—43.
[28] I have explored some of these connections in my *Essays on Wittgenstein and Weininger* ("Studien zur österreichischen Philosophie," Vol. **9**; Amsterdam, 1985).
[29] Lichtenberg, *op. cit.*, p. 88.
[30] Ludwig Wittgenstein, *Philosophische Untersuchungen* (Frankfurt/Main, 19), **I**, p. 122.
[31] *Ibid.*, **II**, xii, p. 230. Lars Hertzberg was the first to point out the importance of fictive natural history in Wittgenstein's thought.
[32] Lichtenberg, *op. cit.*, p. 61.
[33] Ludwig Wittgenstein, *Culture and Value*, ed. G. H. von Wright and Heikki Nygaard, trans., Peter Winch (Chicago, 1980), p. 24, p. 75.
[34] Actually, the author of *Zettel*. Ludwig Wittgenstein, *Zettel*, trans. G. E. M. Anscombe (Oxford, 1967), p. 83.
[35] See Erich F. Podach, *Friedrich Nietzsches Werke des Zusammenbruchs* (Heidelberg, 1961), p. 236.
[36] Lichtenberg, *op. cit.*, p. 121, p. 146, p. 155.
[37] *Ibid.*, p. 54.
[38] *Ibid.*, p. 69.
[39] *Ibid.*, p. 26.
[40] *Ibid.*, p. 120.
[41] *Ibid.*, p. 45.
[42] *Ibid.*, p. 111.
[43] Walter Kaufmann, *Discovering the Mind* (3 vols.; New York, 1982), **II**, pp. 54—7.
[44] Lichtenberg, *op. cit.*, p. 131.
[45] Cited in Stern, *Lichtenberg*, p. 286. His translation.

CHAPTER XIV

TACIT KNOWLEDGE, WORKING LIFE AND SCIENTIFIC METHOD

What is tacit knowledge? What role does it play in working life? What does this imply for our understanding, say, of the implications of new technology such as Expert Systems for working life? What do philosophers of science have to learn from working life studies? These have been the principle issues which preoccupied me in the course of my stay at Arbetslivscentrum. In what follows I propose to answer these questions, at least in a preliminary way, with a view to elucidating the perspective of the project "Utbildning-Arbete-Teknik" with which I have been associated. This amounts to clarifying a number of issues about which there is a good deal of confusion relating to the crucial notion of tacit knowledge, the ways in which theory of knowledge (a term I much dislike but employ for want of a better one) can clarify our understanding of working life, as well as the role working life studies can play in the development of the theory of knowledge. It is, then, at once a compilation of the ideas which have been most central to my work in the Center as well as a statement of what I take back to the academic world from my experiences at the Center. It is hardly intended to be comprehensive; rather it is a sketch of what I, as a philosopher, take to be the most challenging and exciting notions that I have encountered confronting the "real world" of work. My aim, then, is to provide a certain orientation with respect to the role that philosophy can play in helping us better to understand working life but also to suggest how philosophers can learn a great deal about certain vexing problems from working life studies. It is, therefore, important to emphasize that what is presented here is at once a resume and a program rather than an analysis.

With that in mind, we can begin by asking: what is tacit knowledge? In the face of misunderstandings as widespread as they are erroneous, the first thing that has to be said loudly and clearly is that there is nothing whatsoever mysterious or arcane about tacit knowledge. It has nothing whatsoever to do with ESP, Rosicrucianism or any such esoteric lore. It is a familiar part of our everyday experience. Indeed, it is that very familiarity which tends to lead us to overlook its importance when we are dealing with such 'deep' issues such as what thinking or knowing is all about.

What, then, does the term positively cover? There are two senses in which philosophers have come to speak about tacit knowledge. Unfortunately, they have not always been as clearly distinguished as they might have been. It is crucial that they not be confused with one another. The first refers to things we could say if we wanted to but either do not choose to do so or just never get around to putting into words. The second refers to those aspects of human experience which are *wholly* knowable self-reflectively — and, therefore, anything but esoteric — but by their very nature incapable of *precise* articulation. We know things without being able to put what we know into words when, for example, we are able to distinguish the sound of a viola from that of a cello in a musical broadcast, or in the way that an experienced cook knows how to vary a recipe when he discovers suddenly that he does not have exactly what the recipe calls for, as opposed to the way in which we know the time, the date or the temperature.

The first sort of tacit knowledge is the sort that could be articulated but happens not to be. It comprises several species which we can distinguish for the purposes of analysis (in situ they are not so neatly distinguishable). First, there is the sort of thing that is kept secret for political (in the loose sense) or economic reasons. Trade secrets typify this sort of tacit knowledge. Guild masters from time immemorial have been acutely aware of the ways in which their status, power and standard of living often depended upon keeping the tricks of the trade from the ininitiated. We find this concern reflected, for example, in Peter Gullers study of surgical instrument makers.[1] In a similar (but by no means identical) vein medical doctors in the United States have resisted efforts to systematize their methods of prognosis on the grounds that the basis for making claims about a patient's development is a knack that has to be acquired over years.[2] Perhaps this is so, but we shall only be certain of that to the extent that we have ventured to systematize their practice. This sort of problem is reflected in Ingela Josefson's work with the medical profession.[3] Briefly, there are good reasons for thinking that we can say more about this sort of thing than we have said till now, but no less good reasons, as we shall see, for insisting that such expert judgment can by its very nature never be fully articulated in a way that it might be, say, machine reproducible. Be that as it may, one aspect of tacit knowledge has to do with the ways in which elites, such as guild masters, maintain their power by political manipulation of knowledge.

Another sort of tacit knowledge has to do with the sorts of things that people, for whatever reasons have never got around to articulating. It is typical, for example, that nobody pays much attention to the practice of blacksmiths until those practices are in danger of going out of existence. Crisis is usually the motive for articulating things that people have never found interesting or important enough to try to systematize. Many homely skills such as cooking can be passed from generation to generation without anybody getting around to writing down how they proceed. The evolution of cookery book from lists of ingredients with neither precise measurements nor instructions beyond "take milk, flour and eggs beat together, cook in oven till done" to the introduction of precise measurements and step-by-step procedures, often profusely illustrated, exemplifies a sort of tacit knowledge that has found articulation. Many skills incorporate tacit knowledge of this sort which has not as yet been articulated but can be.

A third sort of tacit knowledge which may be articulated, but in fact has not been, has been described with crystal clarity by R. G. Collingwood in Chapter 3 of his *Essay on Metaphysics* under the rubric of "absolute presuppositions".[4] Collingwood spelled out his reasoning there in relation to the inquiries of natural scientists but they are by no means so restricted. In fact, every practice incorporates absolute presuppositions — which by the way are only relatively absolute in that they are absolute relative to a certain enterpize. In fact, Collingwood argues, any effort to answer a question with precision entails making assumptions. The most general and widely held of these are so obvious that it would be silly and tedious to remind ourselves of them continually. It is only when we run into crises, i.e., *wholly unforeseen* trouble that it makes sense to consider them.

An example will help here. We all assume without giving the matter the slightest thought that the money we have in our pockets will be worth something similar tomorrow to what it is worth today. We base not a little of what we do on that assumption. But consider the case of Europe in the early 20s, when sudden enormous inflation hit. All of a sudden one of the rock bottom pillars of European social life crumbled. Thus what is silly and tedious to question in normal circumstances, i.e., what we absolutely presuppose, can suddenly and shockingly be challenged. Then it becomes urgent that we reconsider what we have tacitly been assuming all along. Much of what we 'know' falls into this category. It too is a kind of knowledge that can be articulated when there is a point to doing so. Nevertheless, none of the sorts of tacit knowledge as yet mentioned merits the title in the strict sense, even though we have in fact brushed up against the real thing, because all of the sorts of considerations till now discussed can be articulated if we want or have to do so. However, there is another sort of tacit knowledge which cannot. This is obviously enough the most important type.

There are two sorts of things that defy precise description by their very nature: certain non-visual sensory experience and the procedure involved in following a specific rule. In both cases our inability to describe what we know by no means indicates that our claim to knowledge is dubious. We do not claim that the objects of tacit knowledge are unknowable, for that would be self-defeating: we would be identifying something as something that could not be known. No, the claim is that what is known in the strict sense of being known tacitly cannot be studied scientifically. We are, therefore, talking about the *limits* of science.

Those who have problems with the concept of tacit knowledge tend to identify all knowledge with science and to consider that, as a non-scientific mode of knowledge, tacit knowledge must be defective. But this misses the point massively. In fact the sort of knowledge normally designated as scientific is based upon tacit knowledge, as Michael Polanyi always emphasized.[5]

The first sort of tacit knowledge in the strict sense is sometimes referred to as knowledge by acquaintance or familiarity. It results from sensuous experience. Wittgenstein chose to exemplify this sort of knowledge by contrasting what it is to know and what it is to describe the smell of coffee or the sound of a clarinet.[6] The distinguished English philosopher, Peter Strawson, once posed the question — without being able to say very much that was interesting by way of an answer to it — what would an epistemology that was not based upon the comparison between the mind and the eye, but, say, upon that of mind and ear be like.[7] He is not alone among philosophers in mooting this question. It is Wittgenstein's merit to have illuminated the issue in ways that directly bears upon our theme — though philosophers have for the most part not realized that. Moreover, it is precisely in directing our reflection to the relationship between sensuous experience and description. If we think of the difference between smelling of hearing and describing what we smell or hear, we soon discover that the smell of coffee, the sound of a clarinet, the taste of marzipan, the texture of raw wool or the pain associated with an abcessed tooth, we discover that it is not possible to *begin* to describe any such experiences in such a way that people who had not themselves experienced them could recognize them. The best we could do is to supply metaphors, analogies and hints — "well, its sort of like this . . .". In the absence of these comparisons, metaphors, analogies etc. there is no way to describe what it is to 'know' these things. Moreover, once the people in question have had the requisite experiences for themselves, they shall be satisfied that our inability to describe them is not problematic. The sign of this is that once having had these experiences the demand for explanation ceases to press itself upon us. To have had these experiences and to have reflected upon them — which is what most of us normally, and philosophers *never*, do — is to realize that we do not have to be able to describe them to understand them. The kind of knowledge, then, associated with the ability to make subtle discriminations, such as that between the sound of a clarinet and that of an oboe is rightly described as skill. Here it is important to point out that, while everybody is capable of such judgement (assuming they do not have some special auditory problems), not everybody is *equally* capable of making judgements of the same caliber in these matters. We do not all develop identically with respect to sensory discrimination in much the same way that we do not all grow tall enough to play basketball competitively.

The second strong sense in which we can speak of tacit knowledge depends upon the open-textured character of rule-following behavior. Rule-following activity entails the kind of knowledge that is only acquired through repetition or practice. Learning to play the piano or to compete in the high jump are two cases in point. We must first be drilled in such as way as we come to eliminate obvious mistakes. Only at this stage can we speak of a 'right' and 'wrong' way to play or to jump. Once we have mastered the technique, once we are 'experienced', the idea of a right and wrong way to play ceases to be a significant part of discourse. Indeed, the idea of an algorithm for, say, a piano sonata is as epistemologically absurd as it is aesthetically abhorrent. Mastery of the rules brings with it a freedom to extend them, say, in the way that a Vladimir Horowitz can bring his expertise to bear upon a Chopin polonaise in such a way that we experience it as wholly new. Clearly, here inventiveness is dialectically linked to mastery of conventions.

This, I submit, is what expertise is all about. The ability to master a set of rules in such a way as to be in a position as to extend them when circumstances warrant it. Rule-following activity may originate in rote behavior but it terminates in creative activity. It is essentially creative.

Yet, there is a problem here. It has to do with a widespread misconception with respect to rules, which is deeply rooted both in the popular imagination and in the liberal philosophical tradition from Mill to Rawls and Habermas. I refer to the notion that rules are independent of practices. On this view, we first learn what the rules stipulate, then how to apply them. Kjell S. Johannessen has trenchantly demonstrated just how this puts the cart before the horse.[8] This is not to say that there are no rules which function this way, for there are; however, these rules have the character of norms with respect to some activity we already can perform. They correspond in our musical analogy to playing with a particular style when there are alternatives to choose from. However, this is not the basic sense of rule-following; for we already have to be able to accomplish something to apply these rules. It is the sort of rule-following activity through which we learn how to perform a specific sort of action in the first place — what Kant called constitutive as opposed to regulative ideas — that has to be primary. Here the rule is scarcely to be distinguished from the behavior through which it is constituted. Insofar as the rule can be distinguished at all from the actions through which we stabilize our lives, it is as an *order or command* to repeat a particular action.[9] We are ordered to follow a pattern. No rule need be explicitly formulated, strange as that may seem.[10] It is in fact not strange; for there is as yet no capacity in the child to understand what a rule is. It is through learning to take orders that we come to understand in the first place. At first we learn to carry on mechanically, we soon learn, to our surprize, that we have to take significant variation into account. Basically, rule-following has a peculiar internal order which is determined by our natural history, i.e. by our being the kind of creature we are. It is important to emphasize that what is being claimed here does not challenge any particular theory of child rearing because it is what *all* theories of child rearing *have* to take into consideration. The internal order in learning to follow rules, then, begins with a kind of dressage (*Abrichtung* is Wittgenstein's exact phrase) but terminates in creative activity as we learn that we have to guess how to continue to follow the example we have been given and, ultimately, as we learn to invent new ways of carrying on. Because this procedure is open-ended, mastery of the rules, expertise, consists in being adept at dealing with unforseen circumstances. It is largely a matter of seeing analogies between situations or simply seeing where we are. This activity resists being put into words because there is a very real sense in which we do not know what we do till we have done it.[11] In short, it is an analogical rather than a digital activity. Another way of putting the matter would be to say that there is always and ineliminably the possibility that we can follow the rule in a wholly unforseen way. This could not happen if we had to have an explicit rule to go on from the start. If that was so there would never be the sort of surprizes that we normally associate with discovery. Moreover, if constitutive rules were known before their application, we could never learn to apply rules without more rules. But this is a logical impossibility; for we would have to have a rule to understand how to apply our original rule, another to apply the second and so on to infinity.[12]

It is important that we see what is being claimed here clearly and precisely. It is not that nothing relating to expertise can be put into words. The development of cookery books from the Middle Ages to Fanny Farmer and Mrs. Beaton down to the venerable Julia Child shows that this can be done magnificently. However, precisely because this is a skill upon which we continue to depend and will apply in yet wholly unforseen situations, it cannot be systematized according to the rules of formal logic. The technical reason for this is that the open-textured character of our rules is such that there is no way of determining in advance whether the Principle of the Excluded Middle is always going to hold in the way it currently does. But we should not find this claim daunting; for its obverse is the claim that human skill is something very real and irreducible to rote activity. It is, then, not incompatible with the notion that skill involves an irreducibly tacit element that we urge the skilled to articulate the basis of their activities. The possibility of radical innovation is, however, the logical limit of description. This is what tacit knowledge is all about. To the extent that work is built upon skill tacit knowledge is an essential part of the analysis of its internal order.

How is all of this reflected in the case studies of the Center's "Utbildning-Arbete-Teknik" program? In answering this question I shall restrict myself to examples drawn from Ingela Josefson's studies of medical skill, Peter Gullers study of surgical instrument making in Sweden and Maja-Lisa Perby's work on local weather forecasting.[13]

It is not my aim to recapitulate those studies here but to try to put my finger on the ways in which tacit knowledge turns up as a central element in them.

The first and most obvious point to be made here is that the extent to which tacit knowledge is central to an understanding of working life is tied principally to the extent to which we are discussing *skilled* work. It is, therefore, precisely the extent to which quality in products or services demands skill or expertise that tacit knowledge enter intrinsically into any accurate account of work. The most dramatic way of seeing how this is so comes perhaps from Maja-Lisa Perby's work, which takes its point of departure from questions about decline in competence in local weather forecasting. No small part of the importance of this research lies in the fact that the question of decline in quality of local forecasting has arisen from *within* the community of weather forecasters. Their doubts about the quality of their own work has arisen in the minds of the forecasters from the proliferation of abstract models of the weather with which they have become confronted. Their intuition is that these models force them to work in ways which prevent them from exercising their skills to the fullest extent. In short, their is a clash between abstract models and the concrete practices and procedures in terms of which those models are employed to produce a local weather forecast. Put differently, formalized knowledge comes into conflict with tacit knowledge. Ms. Perby has senstively portrayed the way in which the skill of the local forecaster, like those of the athlete or opera singer require a 'warming up' period before they can be exercised to their full capacity. For the forecasters this means putting themselves into the position where they can synthesize a huge amount of data into an "inner picture of the weather". Tacit knowledge turns up in this process in terms of the role that not only conversations with pilots and colleagues with respect to details about the weather but also the most seemingly casual contacts with them over coffee, say, serve to 'fine tune' their sensitivities. Ms. Perby suggests that the exercise of this sort of intellectual skill, the process of "concretizing" the inner weather picture, is intimately and inextricably bound to sensual skill such that the most precise way of describing the sort of competence involved is to designate it as a mode of *aesthetic sensibility*. Now, this is something which can be described, to the extent that the practices of local forecasters generally permit description, but it cannot be 'analysed' in the sense of being reduced to a series of effective procedures. Interestingly, the same point concerning the aesthetic character of the skill of surgical instrument makers has been stressed by Peter Gullers. It is not the least significant implication of this research that it shows graphically and dramatically that virtually the whole internal structure of such skilled work is lost in a Taylorist analysis.

Ingela Josefson, upon whose research Maja-Lisa Perby to a large extent relies, has emphasized the extent to which medical expertise resists formulation in the form of precise description. This has long been recognized in the area of diagnosis. Physicians have often been able to recognize a patient's condition without being able to say exactly how they drew their conclusions. Ms. Josefson's work with nurses has provided further documentation of the role that such tacit knowledge plays a crucial role in medical care. Here to the ability of nurses to form sound judgments with regard to medical care, corresponding to the doctor's judgment with respect to cure, also turns to great extent upon factors which tend to resist description. As with Maja-Lisa Perby's work with local weather forecasters, Ingela Josefson emphasizes that there is a great deal about the practices of nurses which does permit description. Nurses, too, can say a great deal about what they know at a factual level. It is the process of forming judgments which are accurate that resists precise formulation.

In view of the sorts of considerations which I have drawn from Wittgenstein's later philosophy with respect to tacit knowledge, this is just what we should expect. There is nothing mysterious or implausible about this inability to put what we know into words. Ms. Josefson rightly insists that the problem in our inability to see this lies far more with the mystique of formal education and professionalism in our society than it does it the concept of tacit knowledge. It is an essential element in what we normally term experience that experienced persons have a facility for handling a wide variety of unforseen situations. Just as it is not possible to describe an unforseen situation by definition, it should be reasonable to think that our ability to deal with the situation should not be taken to imply that we *must* be able to articulate the knowledge which enables us to do so. As we have seen, it is Wittgenstein's distinction to have provided an account of why that experience defies precise elucidation. Here, however, the very achievement of these researchers is to show that it is only possible to come to see the problem of describing how expert judgments are formed in the concrete through long and intense discussions in which the researcher comes to understand the perspective of the worker to the point of being able to articulate it. This sort of articulation is best described as a sort of hermeneutic activity; it is a form of interpretation which is dependent upon the researcher's ability to come to grasp the practice of the worker as a participant in those practices might, and not merely as an 'observer'. This sort of hermeneutics is by no means based upon the idea that there is nothing to be said about those practices rather, the very achievements of Maja-Lisa Perby, Ingela Josefson and Peter Gullers consist in having listened sufficiently to workers to be able to describe their activities in such a richly nuanced manner as to lay bare their complexity and, consequently, the competence they require. Thus, we can come to learn why it is that skill rests upon tacit knowledge precisely through a detailed, sympathetic description of agent's practices.

It is not, then, as though one can say "tacit knowledge" like "Open Sesame" and thereby solve a problem instantly; rather, it requires considerable dilligence to be in a position to be able to articulate a situation in working life — and this is crucial — to the point where we can grasp why it eludes precise description. In fact, it is the task of articulating that sort of tacit knowledge which can be put into words but normally is not expressed that brings us to the point where we are able to grasp the limits of description. It is the merit of the case studies mentioned that they do this graphically and excitingly. Moreover, on the basis of the account of tacit knowledge that I began with it should not be surprizing that these studies have been occasioned to a large degree by the efforts of computer scientists to replicate human skill; for the sort of knowledge that these case studies have yielded, as Wittgenstein and Collingwood were fond of pointing out, is only valuable when the certainties upon which the order in everyday life rests are called into question. So, curiously, we have indirectly to thank those very unrealisitic aspirations [14] of computer scientists for motivating researchers to undertake those studies through which we have attained an enriched grasp of the order intrinsic to working life. It remains to look at the significance of Peter Gullers's study of surgical instrument making in Sweden to see the connection between tacit knowledge and yet other important aspects of working life.

The most skilled surgical instrument makers are a very appropriate group to study, since, without being craftsmen themselves, their work requires considerable skill — it is not the sort of work that just anybody can undertake. The importance of this difference is such that it accounts for a certain social differentiation between workers on the basis of their "virtues" — in the Greek sense of *aretē* or excellence. Here it is best to let Peter Gullers speak for himself:

What do these various voices then say about work in a workshop making surgical instruments, and the changes that have taken place over the years? "The old instrument makers" certainly belonged to what we may call an aristocracy of labor, they were regarded — and regarded themselves — as a cut above others who worked in the same shop. Once, long ago, they wore bowlers and stiff collars, and they were as close as could be to craftsmen. This did not mean that they had any great freedom: what was to be done was closely determined, and their security lay in the fact that they were irreplacable. Pay was never particularly high, and hardly higher than other workers in the same workshop got. It was the work itself that was skilled and that gave them prestige rather than anything measured in money terms.[15]

It is only on the basis of such a state of affairs that we can understand, say, Marx's hope that work should be transformed from a necessary evil to life's prime want.[16] Here the aesthetic sensibility involved in producing a quality product is seen to be intrinsically linked to the self-differentiation of workers, and, if you like, the moral world of work.

But what of the implications of the working life perspective for the theory of knowledge? They are profound indeed. I should like to begin by identifying two themes near and dear to Karl Marx that these studies in working life call into question. First, these researches imply that skill remains highly *particular.* The modern (post-modern?) worker is *pace* Marx *not* becoming a universal agent for production. Secondly, and, more importantly, these studies show that to the extent that industries produce quality products or provide quality services, the interests of labor and those of management in maintaining the quality level is identical. To this extent, then, a class-struggle model of labor relations is suicidal. The sooner this is recognized the better. Now to the question of the implications of working life as a perspective for scientific method.

Looking at issues raised within theory of science from the point of view of working life injects a healthy pragmatist humanism into the discussion. The fact that the working life perspective imposes a "real time" framework upon discussions such as those concerning the nature and scope of artificial intelligence, for example, demands that the programs' actual performance, its actual achievements and its impact upon living human beings be placed squarely in the forefront of discussion. The working life perspective rules out the relevance of promises. speculations, thought experiments etc. as irrelevant by placing the issue of performance squarely at the center of the discussion. It forces academic ideologues such as structuralists, behaviorists or cognitivists to one side by requiring that researchers take as their point of departure the practices and procedures of living agents. From there it proceeds to inquire into the impact of development upon concrete individuals. This, of course, by no means that research into these areas becomes easier, only that scientific questions take their sense from human practices, i.e., from knowing as embedded in acting, rather than from speculative *theories* about what it is to know or act. Whatever problems there are with the working life perspective, it serves to wrench us away from the kind of pseudo-sophisitication of armchair speculation — something not incompatible with what is normally, and dubiously, termed 'experimentation' — in an area like psychology, which lead us to ignore the actual circumstances under which real people function. No small part of the significance of the working life perspective for philosophy of science, then, follows from the contextual character of the "real time" point of view which it requires. My claim is that it is a healthy antidote to the sort of metaphysical speculations which dress themselves in the guise of 'experimental' science in areas such as psychology, linguistics or sociology.

This perspective is particularly important for an area such as social philosophy inasmuch as it chastens us to form our expectations for the future in terms of the realities of the present. Thus, from the perspective of a working life researcher like Bjørn Gustavsen the Promethean efforts of Jurgen Habermas to develop a theory of communicative action turn out simply to be irrelevant to labor relations because the stipulations relating to dominance-free dialogue simply do not obtain nor can they be expected to in the forseeable future.[17] Thus, while such a normative theory might conceivably be of some help in judging just what progress we have made, it is too far removed from the actual situation under which communication takes place in working life — from the "essentially contested" character of the concepts involved — to be of much help in facilitating the kind of discussion that must characterize industrial relations in an industrial democracy today. Once more, it is the pressing need to confront the world as it is, rather than as we would like to have it, that is so refreshing with respect to the way the working life perspective illuminates philosophical issues. This

Gustavsen rightly sees as the essential presupposition for effectively introducing change. It does not eschew change by any means but insures that what changes are brought about are cemented into conditions as they actually are, rather than as we would like them to be; or as they 'ought' to be. It is not simply that the working life perspective on communication proceeds from the actual conditions obtaining in a particular situation, it seeks in those conditions the 'logic' or dynamics of change and in that the limits of the possible. Since this is something that philosophers have had great trouble in doing, the suggestion is that they can benefit considerably by looking at traditional questions about social science and social reform from the working life perspective. This suggestion is more radical than it might seem.

One of the most important points in the development of western thought was the point at which Socrates, as the standard story goes, decided to test the Oracle's assertion that he was the wisest of men by 'interviewing', as it were, the other prime candidates for that title: poets, politicians and *craftsmen.*[18] He came way from all with the same disappointment. He came away from them convinced that they did know something or other but that whatever they knew could hardly count as wisdom, since they could not verbalize what they knew. It was the service of Ludwig Wittgenstein to challenge Socrates — and the western philosophical tradition after him — by suggesting that there were in fact good reasons why craftsmen should be considered to possess genuine knowledge and equally good reasons for their inability to provide a generalized or theoretical account of what they know. In all this, it is hardly accidental that Wittgenstein was as committed as Tolstoi to the notion that humanly useful work, from which he tended to exclude the activities of academics, was an integral part of any life that was worth living. As for Socrates, legend has it that the main source of his marital problems was his wife's complaints about his refusal to practice his stonemason's trade! Be that as it may, the problem with Socrates, as with some many subsequent philosophers right down to today, is that he sought the wrong sort of answer to his question. In short, he failed to see that the sort of wisdom that attaches to craftsmanship is tied precisely to the sorts of examples and analogies which he found wanting. So it might be said that the main problem with the western philosophical tradition, starting with Socrates, is its failure to take working life seriously. If this is right, it is not only the concreteness and humane character of the working life perspective — that point at which the Marxian concept of praxis meets Pope Leo XIII's concern for the dignity of labor[19] — that makes working life studies so philosophically impressive, it is the epistemological revolution that they embody. Philosophers ignore that at their peril.

NOTES

* I am grateful to Kjell S. Johannessen for many stimulating and enlightening conversations concerning the philosophical questions discussed here. He is currently preparing an in-depth study of Wittgenstein and Polanyi which will be of great significance for understanding the epistemological issues treated here. I am equally indebted to Bo Göranzon Peter Gullers, Birger Viklund and the other authors cited here for rewarding discussions of working life research, industrial democracy etc. I am not sure to what extent they share the opinions herein expressed. They are certainly in no way responsible for the shortcomings of this paper, which are wholly my own responsibility.

1 Peter Gullers, *Instrument Makers and Surgeons: Some Aspects of Surgical Instrument Making in Sweden* (Stockholm, unpublished).

2 Alvin Feinstein, "Critical Judgment and Basic Science", a lecture to the Boston Colloquium for the Philosophy and History of Science, Nov. 4, 1969 (unpublished).

3 Ingela Josefson, "The Nurse as Engineer", *AI & Society*, Vol. **1** no. 2 (1987), 115—26.

4 R. G. Collingwood, *An Essay on Metaphysics* (Oxford, 1940), Ch. 3, p. 34—48.

5 Micheal Polanyi, *Personal Knowledge* (London, 1973).

6 Ludwig Wittgenstein, *Philosophische Untersuchungen*, **I**, §610; **I**, §78.

7 Peter Strawson, *Einzelding und logisches Subjekt* (Stuttgart, 1972), p. 82ff.

8 Kjell S. Johannessen, "Language, Art and Aesthetic Practice", *Wittgenstein, Aesthetics and Transcendental Philosophy* eds. K. S. Johannessen & T. Nordenstam (Vienna, 1981), p. 108—26.

9 Wittgenstein, *op. cit.*, **I**, §206.

10 Wittgenstein, *Über Gewißheit*, §95.

11 This runs counter to the naturalist account of the matter according to which we do not know what we have done even after we have done it! See below.

12 Wittgenstein, *Philosophische Untersuchungen*, **I**, §§80—8.

13 Maja-Lisa Perby, "Computerization and Skill in Local Weather Forecasting" (Stockholm, unpublished).

14 IBM's claim "get three years experience in one week" is a good example. I owe it to Bo Göranzon.

15 Gullers, *op. cit.*, **VIII**, p. 37.

16 Karl Marx, "Critique of the Gotha Program", *Marx & Engels: Basic Writings on Politics and Philosophy*, ed L. Feuer (Garden City, 1959), p. 119.

17 Bjørn Gustavsen, "Workplace Reform and Democratic Dialogue", *Economic and Industrial Democracy* Vol. **6**, no. 4 (1985), 461—79.

18 Plato, *Apology*, 22c—d.

19 See Pope Leo XIII's encyclical *Rerum novarum*.

CHAPTER XV

MARTIN BRODY AND ALLAN JANIK

PARADIGMS, POLITICS AND PERSUASION: SOCIOLOGICAL ASPECTS OF MUSICAL CONTROVERSY

I. PARADIGMS AND TRANSMITS

Musical composition is unquestionably difficult to discuss apart from the context of the language of technical musical analysis. It is all the more difficult to discuss to the extent that composers become champions of opposing styles. Moreover, we can never be certain that a composer's style (of composition) is consonant with his verbal statements about music. Thus controversies between composers may appear particularly complex and irrational. We wish to reconsider the nature of representative statements of composers and music theorists — both in their informal manifestations (composers criticizing each other's work) and in their supposedly more rigorous applications, i.e., in the interpretive apparatuses of contemporary music theory. We intend to focus our discussion on musical controversy in order to demonstrate one central contention — that controversies in music inherently involve verbal discourse and that *this discourse has a life of its own*, at once independent of and contingent upon the music it describes. Moreover, verbal discourse about music profoundly influences future practitioners of musical composition and shapes our understanding of the music of the past. Of course, a composer may "respond" to a piece of music by composing another piece of music, and these two pieces may be understood as representative of "competitive aesthetics". But even to say this is to move from musical to verbal discourse. We wish to explore some of the senses in which this verbal discourse is shaped by social forces which partially determine its "closeness of fit" to the objects it is meant to describe, that is, musical compositions. The relation between musical objects and the language of musical analysis is not innocent. Analytical discourse is inherently polemical, and it develops in relation to other analytic discourse, both within and outside of music, as well as in relation to the notation and experience of music itself. In the discussion which follows, we hope to illuminate these assertions through comparison of scientific and musical controversies. These comparisons are not meant to constitute a fully developed sociology of musical knowledge. Rather, we wish to propose a group of questions which may suggest some new ways of approaching the problem of discourse about a non-verbal art.

A possible approach to the description of musical controversies is drawn from the language of philosophy of science and specifically from the vocabulary of Thomas Kuhn. Commentators have been inclined to speak of controversies concerning musical style or analytic method as clashes of "paradigms".[1] However, if we look carefully at what Kuhn means by paradigm in his *Structure of Scientific Revolutions*,[2] it seems that this apparently useful transference from science to music is not desirable. We wish to explain why the concept of paradigm clash is inappropriate to the discussion of musical composition and also to suggest that there is another terminology available, also drawn from current discussions of philosophy of science, which is illuminating when brought to bear upon controversies in music. Thus, we will begin with a general discussion of the Kuhnian notion of paradigm with a view to exposing the problems involved in transferring the notion from science to music. We will proceed to an alternative model of transmission of knowledge for diverse disciplinary types and consider the relevance of this model to music. Finally, we will present relatively extensive examples of scientific and musical controversies and the different modes of resolving controversies characteristic of the different disciplines to be discussed.

The ensuing discussion may seem to be highly abstract, resting upon speculative assertions and apparently invidious comparisons. Thus, we shall attempt to articulate some of the interests and problems which led us to collaborate on this topic.

Our interest in this topic emerged from discussions in an informal seminar in which Kuhn's paradigm language began to recur as a way of expressing the kinds of clashes that exist between twentieth century "schools" of composition. As a philosopher and intellectual historian, Allan Janik had long been interested in the use and abuse of Kuhn's language; he raised the question as to whether the people who used this expression actually alluded to Kuhn's use or were simply employing the term as a fashionable cliche. It emerged that typical conversational usages of the term paradigm (the kinds of uses we are criticizing only rarely find their way into print but are no less important for that reason, for in our opinion they reflect the type of beliefs that are deeply held but seldom articulated) were often closure moves in conversation. Thus, to speak of a musical controversy as a clash of paradigms is to assert that there are a number of ways of composing, analyzing music and discussing it generally, each of which is incomparable to the others, each of which is intended to be definitive and exclusive. Talking about musical controversies this way amounts to endorsing a conception whereby diverse but highly organized schools of composition and music theory operate with fully articulated goals and procedures. The goals and procedures of different schools would be, for the most part, distinct. Thus, such an appeal to clashes of paradigms seems to conceal the nature of musical controversies by preventing the crucial step of proceeding to produce "thick descriptions" of what actually transpires in interchanges between schools of thought. Moreover, the closure move bars the way to substantive questions of the form "is this controversy really about the issues the antagonists claim it is?". That is, it bars the way to questions about "false consciousness" among participants.

From the side of music theory and composition, Martin Brody has been concerned with the interaction of the two activities, that is, the way in which theoretical nomenclature may shape the concepts which composers invoke in the compositional process. Similarly, the question of how a new compositional achievement may undermine publically accepted, reified analytic procedures must be raised. A formalist account of music may fail to reveal the subtle and, in some cases, continual modifications of concepts used by professionals to explain and evaluate compositional achievements.

We are very much aware that there are myriad difficulties and lacunae in what follows but we are not less certain that we have struck upon a scheme for looking at musical debates with a definite heuristic value. We have to emphasize that we are not advancing anything like a method but aim only at introducing a perspective on music. To that end we are far more interested in introducing the right questions than any dogmatic pronouncements about how musical controversy *must* be understood.

To begin, we ought to bear in mind that Kuhn introduced the notion of paradigm to explain scientific growth or development in terms of radical changes in scientific conceptual apparatus and the scientist's world view. Margaret Masterman has identified over twenty senses in which Kuhn used the term paradigm in the *Structure of Scientific Revolutions*.[3] It would be tedious to enumerate all of them here, but it will be of interest to look at the range of things, the multitude of sins, as it were, that fall under its domain. According to Masterman's account, Kuhn uses the term paradigm *inter alia* to designate: a universally recognized scientific achievement, a myth, a "philosophy" or constellation of questions, a textbook or classic work, an accepted device in common law, an analogy, a standard illustration, a gestalt figure which can be seen in two ways, a set of political institutions, a whole tradition in science. The sheer diversity of these senses of paradigm in itself should be a warning that the term is mightily ambiguous and probably questionable in its own context, let alone when transferred to another one. It is worth noting in passing that paradigm has become something of a buzz-word in intellectual circles; thus, the phenomenon which concerns us in our discussion of music is not entirely isolated from other discussions of paradigms and paradigm clashes concerning areas as diverse as economic growth or sex roles.[4] Hence, these comments may have consequences beyond our immediate area of concern.

Of the many connotations of the term paradigm, three at least are crucial to Kuhn's argument and to an understanding of what the notion aims at representing. First, paradigms are *analogies* or *metaphors* which make discovery possible, i.e., allow us to map new territory scientifically. Thus, by considering electricity as a fluid, we make it possible to discuss the "flow" of "current" in such a way that "flow" can be measured with precision. Second, paradigms are theories inasmuch as they become models for future theorizing. For example, Galileo's theory of how freely falling bodies accelerate became a model for the development of physical theory generally. Third, paradigms are sets of problem solving procedures which incorporate the acknowledged standards of the scientific community and are thus capable of being passed on from generation to generation of scientists. They play a role in the process of discovery, the logic of theory development, and the sociology of disciplines.

As we shall see, these senses of paradigm are pertinent to the understanding of changes in and transmission of musical knowledge. However, another crucial feature of paradigms poses problems for musical applications. It is very important for Kuhn that sciences can be pre-paradigmatic or post-paradigmatic. The early history of chemistry is an example of a pre-paradigmatic stage of scientific development, when properly chemical and alchemical notions are mingled helter-skelter. The emergence of the paradigm makes the science of chemistry possible. Kuhn characterizes this situation curiously as one in which there are scientists but nothing which can properly be called science. Something similar happens when a discipline ceases to be paradigmatic. This happened to philosophy, according to Kuhn, with the dissolution of Scholasticism and to painting with the passing of the idea that art should strive at exact representation of nature. Thus the contemporary situation in, for example, philosophy (not on the face of it unlike that of music, in that different conceptions of what it means to do philosophy compete with one another) is not one in which the discipline can be said to be paradigmatic. It is important to note that the kinds of situations in which it might be tempting to discuss clashes of paradigms in music are situations in which Kuhn suggests that paradigms are precisely what is lacking. The clash of paradigms, it should be remembered, is something that historians of science can analyze only after the fact. (The Owl of Minerva takes wing only at dusk.)

The widespread use of the term paradigm has prompted not a little suspicion concerning its capacity to obfuscate rather than clarify the myriad issues into which the notion has been introjected. This may well have a psychological origin, since, apart from students of classical languages who will remember committing to memory the models for conjugating verbs and declining nouns, no image of a "paradigm" comes readily to mind when we hear this word. Clearly, it we substitute Imre Lakatos' equivalent, research program,[5] for the term paradigm, we would probably be, on the whole, less inclined to employ the idea generally. However, the distinctions between these terms is important. Understanding scientific development involves a clarification of the relationship between the concepts of science, research program, discipline, scientific analogy, etc.

Stephen Toulmin has made inroads in understanding the nature of disciplines by focussing on the notion of what a discipline transmits. We shall see that Toulmin's notion of a science as organized around what can be transmitted over generations has much in common with Kuhn's but is significantly less ambiguous. Moreover, Toulmin's model is inherently comparative. It provides criteria for classifying types of disciplines in relation to a model of the transmission of knowledge in science. In short, the "transmit" language of Toulmin is more fully articulated than the "paradigm" language of Kuhn. What, then, are transmits?

For knowledge to be publicly transmitted we require (1) an unambiguous technical vocabulary (2) a technique for representing nature which is based upon certain ideals of natural order (3) agreed upon procedures for applying our representational techniques, i.e., some sense of the procedures which allow us to use these techniques in experimental situations. Not all disciplines, however, are equally scientific. For disciplines vary as to the extent to which their *problems are well-defined.* Toulmin proposes the following criteria for determining whether a given discipline's problems are so structured, or in his language, "compact":

> A rational enterprise whose conceptual repertory is exposed at every stage to critical reappraisal and modification by qualified judges in the light of clearly recognized and agreed collective ideals [is compact].[6]

Toulmin goes on to assert that these disciplines possess five connected features:

(1) The activities involved are organized around and directed toward a specific and realistic set of agreed collective ideals.

(2) These collective ideals impose corresponding demands on all who commit themselves to the professional pursuit of the activities concerned.

(3) The resulting discussions provide disciplinary loci for the production of "reasons" in the context of justificatory arguments whose function is to show how far procedural innovations measure up to these collective demands . . .

(4) For this purpose professional forums are developed, within which recognized "reason-producing" procedures are employed to justify the collective acceptance of novel procedures.

(5) Finally, the same collective ideals determine the criteria of adequacy by appeal to which the arguments produced in support of those innovations are judged.[7]

Toulmin recognizes that there are some disciplines that approximate these criteria, i.e. those in which there is indeed a transmit (collective ideals and procedures) and a collective forum but which do not have specific collective problems and/or goals. Musical composition and poetry fall into this category. In these areas there is a sufficient, professionally-accepted technical component to form a transmit even in the absence of common goals on the part of the individuals who utilize those techniques to form a disciplinary core. Toulmin refers to this second class of disciplines as "diffuse" as opposed to "compact" disciplines. (His third category comprises such subjects as anthropology and psychology in which there is no recognized transmittable core. Toulmin terms these subjects "would-be" disciplines.)[8] Music has a technical vocabulary shared by professional practitioners (though it is, as we shall see, hardly unambiguous). Moreover, the disciplines of music share inter-related models for generating aspects of analyses (for the theorist) and form and syntax (for the composer). Aspects of these models will be considered in the examples below. What is often missing in musical composition and theory is consensus about common ideals, which, Toulmin asserts, distinguishes compact and diffuse disciplines. Thus, "arguments produced in support of . . . innovations" may be particularly controversial.

It is our intention to explore some of the implications of Toulmin's thinking by contrasting certain selected aspects of competition between scientists with competition between composers and music theorists; we hope to clarify the suggestion that musical composition and music theory are "diffuse" disciplines. We have focused upon competition because it ought to highlight the differences between disciplines organized around common goals and those whose goals are determined by individual practitioners. At the very least we shall be able to note interesting parallels between scientific and musical competitions, such that we may see musical competitions as something other than aberrant phenomena. However, before launching into that discussion we shall have to add our own wrinkle to Toulmin's compact/diffuse distinction by supplementing his account with material drawn from current discussions about the methodology of science; but here we shall be drawing upon the philosophy of social rather than natural science.

II. ESSENTIALLY CONTESTED CONCEPTS

We believe that the literature arising from William Gallie's discussion of *essentially contested concepts*, specifically the development of that notion in William Connolly's penetrating study, *The Terms of Political*

Discourse, will be of value in further unpacking Toulmin's distinction. Gallie defines essential contestedness as follows:

In order to count as essentially contested . . . a concept must possess the following four characteristics: (1) it must be *appraisive* in the sense that it signifies or accredits some kind of valued achievement. (2) This achievement must be of an internally complex character, for all that its worth is attributed to it as a whole. (3) Any explanation of its worth must therefore include reference to the respective contributions of its various parts or features; yet prior to experimentation there is nothing absurd or contradictory in any one of a number of possible rival descriptions of its total worth, one such description setting its component parts or features in one order of importance, a second setting them in a second order and so on. In fine, the accredited achievement is initially variously describable. (4) The accredited achievement must be of a kind that admits of considerable modification in the light of changing circumstances; and such modification cannot be prescribed or predicted in advance. For convenience I shall call the concept of any such achievement 'open' in character.[9]

To which Gallie adds, (5) "each party recognizes the fact that its own use of it is contested by those of other parties . . . to use an essentially contested concept is to use it aggressively and defensively".[10] It should be tolerably clear that the more compact a discipline the fewer essentially contested issues it will contain, for a compact discipline will be characterized by shared goals and criteria. In a "would-be" discipline nearly every concept will be essentially contested. Diffuse disciplines will be characterized as those in which concepts are both shared and contested; and in which methodological reflection will at least in part take the form of distinguishing what is really shared from what is really contested (as opposed to what merely seems to be shared and contested). Clearly many of the concepts relating to musical composition are essentially contested according to Gallie's criteria. Thus, we should not expect musical controversies to be casual and occasional but essential asspects of the diffuseness of discipline. We shall illustrate this point below.

William Connolly's work has important implications for our program because he has introduced several corollaries to Gallie's thesis. First, Connolly suggests that essentially contested concepts (e.g. politics, art, religion) should be considered *cluster* concepts. In the case of the concept of politics, his immediate concern, he writes:

To make the concept of politics intelligible we must display its complex connections with a host of other concepts to which it is related; clarification of the concept of politics thereby involves the elaboration of the broader conceptual system in which it is implicated.[11]

The second of Connolly's corollaries follows closely from the first; it is the notion that the distinction between analytic and synthetic statements is problematic in connection with cluster concepts:

> The analytic-synthetic dichotomy itself breaks down when we confront cluster concepts such as politics . . . If, in elucidating cluster concepts, the analytic-synthetic dichotomy breaks down, modes of analysis for settling conceptual disputes that pre-suppose that dichotomy also loose much of their purchase . . . At the most general level, the break down of the analytic-synthetic dichotomy diminishes the utility of formal methods of logical analysis in clarifying cluster concepts.[12]

The third corollary Connolly draws from Gallie's thesis follows similarly from the second. It is the notion that cluster concepts can never be exclusively descriptive but are always normative as well. He writes:

> If we subtracted the moral [or normative] point from any of these concepts, we would subtract as well the rationale for grouping the ingredients of each together within the rubric of one concept . . . And if we were to exorcise the evaluative point from any of these concepts, we would be at a loss as to how to clarify or refine its boundaries when new and unforseen situations arose . . . the dialectical relation between the criteria of a concept and its point or purpose in our language is exactly what makes the notions 'democracy' and 'politics' the subject of intense dispute.[13]

We shall see that the distinction between "criteria" and "point" is equally pertinent to many concepts in music theory and composition. The point of introducing these notions here is that they serve to complement Toulmin's development of the idea of a diffuse discipline as one in which techniques are shared but ideals differ; Connolly's arguments help to explain how a diffuse discipline is conceptually possible. Changes in clustering, differing estimates of what is analytic and what is synthetic and consideration of differing normative points of veiw will all help to elucidate the specific "mode of diffusion" within a diffuse discipline. The musical illustrations below will, hopefully, flesh out this discussion. Let us now characterize competition in the "hard sciences".

III. AN EXAMPLE FROM SCIENCE: THE "RACE" FOR DNA

The most striking feature of modern scientific competitions is that they terminate in victories which have been described as *absolute* and *substantive* (i.e., either you have or have not made a discovery, either you are first or not). This helps to explain why competitions between scientists typically occur in the initial stages of research, i.e., at a point when it is unclear who is most likely to solve the problem in question. Temporal priority is all important. Thus, the sense of being involved in a race is almost inevitable. The theoretical breakthrough which constitutes victory produces a complex unity of elements hitherto unthinkable and, thus, by relating two or more natural phenomena, opens the door to further research by specifying *precisely* what it is that we currently do not know. When one of the competing research programs succeeds in establishing what is actually the case, the other(s) become more or less superfluous and must go out of existence or be relegated to a very secondary type of research. In what follows, we have chosen the case of the "race" to discover the structure of DNA to illustrate some aspects of scientific competition because, drama aside, it is perhaps the most widely known and accessible case in the recent history of science. In the case of the competition to discover the structure of DNA, the method of searching for equations to fit x-ray diffraction patterns practiced by Wilkins and Franklin became superfluous once Watson and Crick succeeded in building a model of the molecular structure of DNA. Watson and Crick won and their competitors lost; it was as simple as that. In this situation, temporal priority was all that mattered; the merits of the alternative approach ceased to be interesting.

A second feature of scientific competition which emerges from an examination of the controversy over the discovery of the structure of the DNA molecule is the tension between competition and communication in science. In her discussion of the role of competition in science, which forms part of Anne Sayre's effort to rehabilitate Rosalind Franklin, Sayre makes the following assertion: "Rivalry is stimulating and useful, and thus is the very way in which science works".[14] The majority of philosophers of science today would find nothing objectionable in that assertion. It is imperative, however, to distinguish, as Sayre and most sociologists of science do, two senses in which scientists can be in competition, for one is extraneous and, in all likelihood, detrimental to the search for knowledge, whereas the other is of its very essence. Scientific rivalries can be rooted in personal ambition — modern scientists engaged in truly original research are always acutely conscious that they are competitors for the Nobel Prize — or it can be rooted in a confrontation of different ideas and methods of approaching a given problem. Both of these factors spur on rivalry and whet the scientist's appetite for victory — but also motivate the scientist to take risks.

On the other side of the picture, the need to communicate functions as a break on scientific rivalry, for, as Toulmin and others have pointed out, science is essentially *public knowledge*. There is no scientific achievement unless every aspect of the experimentation and theorizing has been subjected to critical scrutiny on the part of qualified judges. There can be no private achievements in science and, for that reason, the scientific community must be dedicated to fair play as limiting rivalry. Trust, therefore, is a value which is essential to the scientific enterprise.

This brings us to a third feature of modern science which is important to our discussion; namely, the notion that modern research requires a high degree of organization.[15] Races in science are not typically between individuals but between laboratories. The DNA controversy provides us as laymen with valuable insights into the workings of science. For instance, one cannot escape the fact that it would have been virtually impossible for any single individual to come up with a solution to the immensely complex problem of DNA, whatever account of the matter one chooses to endorse. Actually, a glance at almost any scientific periodical will serve a similar purpose in that its table of contents will typically include a very large number of co-authored pieces, certainly far more than are typical of other areas of intellectual endeavor.

Finally, we should emphasize that the very notion of a race between scientists is ambiguous and problematic. We have seen that "race", in this context, can refer to a contest between individual scientists or a competition between ideas and techniques. It might also be both of these things. The point of emphasizing this is that the very notion of what the rivalry is all about may well be a matter of controversy.

In the case of the DNA controversy, the notion of a "race for DNA" only appeared after the fact in Watson's account. Watson emphasized his fears that Linus Pauling at Cal. Tech. would discover the structure of DNA before Crick and him at the Cavendish Laboratory at Cambridge. Watson's expressions in describing various developments on the path to the discovery of DNA are worth noting for the tense character they lend to his narrative. He calls Pauling's "capture"[16] of the alpha helix a "bleak situation",[17] as the thought of competing with the Nobel laureate, Pauling, dawned on him as well as the fear of being "humiliated"[18] at the loss of the prize. Pauling had already solved the problem of the structure of the alpha helix, one of the first breakthroughs in determining the structure of a biologically important macromolecule. According to Watson, this was one of the bases for modeling the ways in which molecular bonding might occur (rather than producing mathematical equations which would account for x-ray diffraction patterns, as Wilkins and Franklin were doing at King's College, London). Watson's description, on the other hand, minimizes the possibility that Maurice Wilkins and Rosalind Franklin, Wilkins' research assistant according to Watson, might have been well on the way to discovering the structure of DNA. One of the main reasons for this, according to Watson, was the degree to which Rosalind Franklin's ill-temper not only made collaboration impossible but actually transformed Wilkins' life into an "emotion hell".[19] Wilkins is portrayed by Watson as anxious but afraid to fire the "less than attractive feminist", whose intellectual abilities could not be denied.[20] Franklin's role, according to Watson, was secondary. The tension in Watson's gripping *The Double Helix* comes almost entirely from California, not from London. It is a contest between the youthful innovator and the old master. To the extent that Watson portrays the race for DNA as a race between methods, however, it is a contest between atomic model building, a common-sensical task of asking "which atoms like to sit next to each other", and recherché mathematical reasoning about x-ray diffraction pictures.

Sayre, a personal friend of Rosalind Franklin and wife of a distinguished scientist, has challenged all of this, arguing that, on Watson's account, there was never any real danger that Pauling was on the right track. She insists that Pauling is a "red herring" in the plot of *The Double Helix*, that the real function of the book, apart from recording his path, was to rationalize his shoddy treatment of Rosalind Franklin.[21] For Sayre, Watson had a create a myth around "Rosy" as well as a false sense of competition with Pauling in order to mask both his unacknowledged borrowings from her work and to emphasize his victory over a Nobel laureate, a much more impressive thing than a victory over the lesser-known Franklin. Sayre emphasizes that Franklin was never an assistant to Wilkins but an equal throughout. Moreover, Sayre attributes Franklin's inability to communicate with Wilkins to a vast difference in scientific style; the former was prone to "a kind of hot and heavy form of disputation" which the latter could not abide. Further, Sayre disputes Watson's allegation that Franklin was opposed to the idea that the structure of DNA was helical. Sayre insists that, despite Watson's suggestions to the contrary, Franklin's x-ray diffraction photographs were absolutely necessary to the solution of the problem. She alleges that Watson's discounting of the mathematical approach that Franklin was pursuing is part of a huge rationalization on his part. It is not necessary to agree with this view (though it has much to recommend it) to see that the kinds of things that scientists say about science are as questionable ideologically as the assertions of the practitioners of other disciplines. The value of Sayre's work is to sensitize us to the fact that first hand accounts of scientific competitions have to be taken with a grain of salt, just as they must in other areas of endeavor. Our point is that the description of scientific controversies can be expected itself to be controversial.

Having presented the notions of compact and diffuse disciplines and an example of competition in a compact discipline, we must proceed to three tasks. First, we have to articulate the questions about music that emerge from our discussion of scientific competitions. Second, we have to look at some typical examples of competitions between composers and music theorists, with a view to establishing some parallels with our example from science. Finally, we intend to use these examples to answer some of the questions we shall have raised.

What kinds of questions, then, emerge from our sketch? How can we use a comparative study of competition in science and music to enlighten us about the nature of the controversies which surround music? Here are some of the possibilities.

First, there are questions surrounding the nature of victory in musical competitions. What is it to win in a music controversy? If victory is not absolute and substantive, as it is in science, how shall we characterize it? If there are not such triumphs in music, does this imply that there is something valuable in the very existence of controversy? If a scientific triumph allows us to ask further questions of nature, what does a musical breakthrough accomplish? What constitutes growth in musical knowledge or technique? How important is temporal priority in musical innovation?

Secondly, there are questions about the relationship of the individual practitioner to the research group or "school" to which he belongs. How does competition between schools compare to competition between individuals? Does rivalry between composers move them to be bolder? Does rivalry serve a disciplinary function, i.e., aid in the growth of the discipline?

Thirdly, there is the question of the relationship between musical composition (what gets written on the page) and the programmatic statements or manifestos of composers or theorists of music. What do these manifestos illuminate? What do they conceal? How does the context in which the composer operates effect his music?

Finally, to what extent is musical composition "collectivized"? If the functions of "good reasons" for innovation, the forums for presenting and defending innovation, etc., are not collective, then what is their actual function? We shall not be able to deal with all of these issues in all their complexity, but we are convinced that the effort to do so will illuminate our subject.

IV. CONTROVERSIES IN MUSICAL COMPOSITION: SEVERAL VIEWS OF NEO-CLASSICISM

The vividness of musical controversies is unquestionable, whether such controversies arise from the competition between contemporary composers, clashes of alternative historical and/or theoretical approaches to the interpretation of music, or simply from unexamined discrepancies in taste. On whatever level, such competitions and the musical accomplishments which consequently emerge, are rarely described in the language of discovery, invention or justification. We wish to present two examples — one a reading of a well-documented controversy (between Stravinsky and Schoenberg), the other, a hypothetical dialogue between music theorists — to exemplify several aspects of musical competition.

The controversy between modes of musical neo-classicism in the 1920's — that is, the controversy between Schoenberg and Stravinsky, or various enthusiasts of the "schools" they represented — is a significant instance of musical competition. It is clearly more important (albeit more difficult to characterize) than the apparent competition between Schoenberg and Hauer, during roughly the same period, over the "discovery" of the twelve-tone system. The case of Schoenberg and Hauer demonstrates the inadequacy of the language of discovery to account for musical innovation; for one thing, Schoenberg and Hauer quarrelled retrospectively over non-commensurate discoveries — their versions of twelve-tone theory differed both in substance and application. Detailed consideration of the composition or theory of J. M. Hauer is beyond the scope of this paper. Suffice it to say that Hauer and Schoenberg quarrelled bitterly over the apparently honorific title "inventor of the twelve-tone system". The similarities in their formulation of the twelve-tone theory derive from a mutual interest in the premise that the aggregate (the twelve pitch class set) could be the essential pitch-structural unit in contemporary music. In its details Hauer's system focussed on a division of the twelve pitch class set into two halves, without regard to their internal ordering. Schoenberg's system involved much greater preoccupations with ordering the twelve pitch classes, and the implications of such ordering. Thus the basic unit for Hauer was the "trope", i.e., the unordered hexachord, and for Schoenberg was the row, i.e., a particular ordering of the twelve pitch class set.[22] Although the work of Hauer did stimulate Schoenberg to present publically some of his initial research before he intended, neither Hauer nor Schoenberg could be said to "win the race to twelve-tone composition". Victory, whatever that might mean in this case, is substantive but not absolute — for the theoretical victory does not preclude parallel victories of competitive systems, each of which might generate (or be generated out of) musical compositions. The apparent failure of Hauer is not so much due to considerations of temporal priority so much as the rejection of the compositional applications of his theory by the community of composers, theorists and historians of music.

The case of Hauer and Schoenberg demonstrates the possibility for a substantive, but not absolute victory in musical compositional controversies. in fact, interest in Hauer seems to be increasing and a small-scale controversy about his importance has resulted.[23] As we have suggested, Hauer's theory of twelve-tone music was in direct, substantive conflict with Schoenberg's, and, in a sense, Schoenberg won — his approach was far more influential than Hauer's. Only Schoenberg's theory "passed the empirical test". But this victory was not absolute, precisely because it was possible for Hauer's compositional methods to re-emerge later and contribute to a body of theory and practice which co-exists with Schoenberg's despite the discrepancies between the two. In short, controversies between music theorists or composers may terminate with several possible outcomes: (1) one of the two participants may win substantively and absolutely, i.e., one set of compositional ideals may be accepted in the professional community while the other appears to be permanently rejected. (2) In contrast, both may be rejected or (3) both may be accepted: controversy may be resolved by a sort of disciplinary bifurcation, whereby both sets of ideals (which may have common dimensions) may emerge. (4) Finally, as in the case of

Schoenberg and Hauer, both protagonists may be accepted, but to different extents and at different times. Clearly, conflicting ideals may exists simultaneously within the discipline of musical composition and theory. Thus, the notion of *paradigm* (and its corollary, "normal science") can only be deceptive in application to music. We are reminded of Gallie's requirement for an essentially contested concept, that "the accredited achievement must be of a kind that admits of considerable modification in the light of changing circumstances".[24] We do not mean to suggest that transmittible ideals are not possible in music. We will elaborate on these observations in the examples which follow.

Despite the sense of controversy that their music aroused, no appraisal or rejection currently surrounds the music of either Schoenberg or Stravinsky. Although Adorno contributed the most monumental testament to the Schoenberg-Stravinsky controversy[25] the protagonists themselves provided substantial documentation of competition:

> It may be appropriate to build exhibition buildings to be pulled down after three months; even pyramids need not be planned for all eternity. But, on the other hand, the aim in seeking a cancer cure can only be a permanently effective one; we want to know something unchangingly valid about the course of the stars and the fate of the soul after death.
>
> Maybe for Stravinsky art falls not into this last category but among fashionable materials and neckties. In that case, he is right in trying merely to satisfy the customers.
>
> Arnold Schoenberg ("Igor Stravinsky: *Der Restaurateur*", 1926)[26]

> Why who could be drumming away there?
> If it isn't little Modernsky?
> He's had his pigtails cut.
> Looks pretty good!
> What authentic false hair!
> Like a peruke!
> Quite (as little Modernsky conceives of him)
> Quite the Papa Bach
> (text from Schoenberg's *Three Satires* for mixed chorus, op. 28, 1925)
>
> I do not know what I am supposed to like in *Oedipus*. At least it is all negative: unusual theatre, unusual setting, unusual resolution of the action, unusual vocal writing, unusual acting, unusual melody, unusual harmony, unusual counterpoint, unusual instrumentation — all this is 'un' without being *anything* in particular.
>
> Schoenberg, ("Stravinsky's *Oedipus*", 1928)[27]

In this first quotation, Stravinsky is the propagator of all things temporary and inorganic — neck-ties vs. cancer cures and the fate of the soul. In the first two citations, Stravinsky is associated with disembodied style, literally in the *Satires* text, in which he is made out to be a character escaped from Boucher — Cherubino scalped.

These attacks are curiously detached from specific musical examples (which abound in Schoenberg's articles on Viennese music) and are diffuse, even confused in the pattern of their imagery. There is a vague suggestion of incongruity between formal and structural properties in Stravinsky's music, but this is an all but hidden sub-text.

For his part, Stravinsky could be equally histrionic in his references to Schoenberg. Thus, he dismissed Schoenberg's *Pierrot Lunaire* in his autobiography, written in 1935:

> I did not feel the slightest enthusiasm about the esthetics of the work, which appeared to me to be a retrogression to the out-of-date Beardsley cult. But, on the other hand, I consider that the merits of the instrumentation are beyond dispute.[28]

The reference to instrumentation amounts to extraordinarily faint praise, considering the unprecedented harmonic and formal properties of Pierrot. Other musicians closely associated with Stravinsky were more explicit in their criticism of the "Schoenberg aesthetic":

> To treat the matter dialectically, Schoenberg may be considered the *Thesis* and Stravinsky the *Antithesis*. Schoenberg's thesis is an egocentric conception dominated by personal and esthetic elements which assume the significance of a fetish . . . Stravinsky's whole aim, on the other hand, is to overcome the temptations of fetishism in art, as well as the individualistic conception of a self-imposed esthetic principle. . . . The principle here affirmed is the limitation of the ego and its subordination to superior and eternal values.
>
> Arthur Lourie, ("Neogothic and Neoclassic", 1928)[29]

Controversies between composers, pertaining to the transmission of musical values, are often similarly broad in their arguments. Allusions to the eternal or perhaps a non-metaphysical, "extra-musical" essence (perhaps to other arts or even politics) is invoked. The argument is metaphorical: appeal is made to some external discipline in which truth criteria or some apparent, transmittible standards of evaluation are thought to exist. Thus, the competition between Schoenberg and Stravinsky resembles the race for DNA in a least some senses. The composers, perhaps even more than the scientists discussed previously, call up elaborate metaphors to enhance the sense of competition between them. In our discussion of the DNA controversy, we argued that scientists in competition failed to acknowledge mutual influence.

The composers of our example do not refer to their competition as a *race* (though Schoenberg's student, Webern, seemed to approve of such a conceit, as the title — and contents — of his pamphlet "The Path To New Music" suggest). Nonetheless, Schoenberg and Stravinsky's verbal statements may be taken to suggest that the discipline was not big enough for two independent sets of ideals. Moreover, they appear to have ignored any correspondences of technical procedures between their works. Both composers encourage the growth of compositional schools based on their ideals, thus stimulating achievement parallel to their own and suggesting the priority of their ideals by professional concensus. Dualisms (fashion vs. nature, eternal value vs. fetish, "neogothic" vs. "neoclassic") are invoked to support the claims of composers and their schools to exclusive authority. The sense of competition seems to stimulate creativity, but it also seems to be detached (or at least detachable) from considerations of transmittible compositional techniques.

Clearly, the controversy between Schoenberg and Stravinsky generated diverse appraisals and descriptions of their achievements. This diversity was possible because of the internal complexity of the achievement being described; this internal complexity allows for different orders of importance among the constituent parts of the musical achievement. These different orderings are invoked by rivals in the appraisal of the composition. In short, the ideals of the competitive schools are expressions of an essentially contested concept, here that of musical composition itself. This essential conceptual contestibility is an important characteristic of diffuse disciplines.

We have already referred to the supporters of Schoenberg and Stravinsky as members of compositional schools. It is important to note that these "schools" are extraordinarily different from scientific laboratories. There are no public, predetermined procedures for research in compositional schools, nor are there any inherent demands for eventual publication of the procedures which develop. In fact, the members of the Schoenberg school were inclined to secrecy — Any revelations of working methods were involuntary and often informal, not based on any clear demand created by the external requirements of music as a discipline. Thus, the tension between competition and communication during (or about) the compositional process is not legislated by composers, the practitioners of the discipline.

V. MORE RECENT VIEWS OF NEOCLASSICISM: CHANGING THE TERMS OF DISCOURSE

If we attempt to seek "more technical", perhaps more intrinsic discrepancies of procedure between musical examples of Schoenberg and Stravinsky, we will notice that the same qualities of essentially contested concepts noted above still emerge. The internal complexity of the achievements of Schoenberg and Stravinsky admits multiple descriptions in which different compositional elements may be differently evaluated. Of course, the language of atonal music analysis has continually changed since the period of controversy between Schoenberg and Stravinsky discussed above. This change has occurred gradually, without concensus among professional musicians. We do not intend to document these changes in detail. A couple of examples may indicate something of their nature. Let us turn to examples from the music itself (fragments from the "Menuett" from the Schoenberg *Suite*, op. 25, 1921 and the variations movement of the Stravinsky *Octet*, 1922—23) to determine whether more detailed technical aspects of musical controversy may be observed.

Certain obvious differences may be noted: the Schoenberg "Menuett" preserves aspects of classical minuet phrasing and form, but it seems to bear less surface resemblance to the classical dance model than Stravinsky's waltz bears to its models. The Stravinsky example contains more conspicuous remnants of tonal music, particularly the accompanimental triad, usually presented in every other pair of measures. However, much of the nomenclature and many of the procedures of contemporary music theory suggest underlying similarities between these two examples.

In this discussion of the Schoenberg excerpt, Seymour Shifrin calls attention to linear and harmonic patterns emerging from the twelve tone structure of the movement, contributing to the articulation of subtle phraseological effects:

It is Schoenberg at his most overtly neoclassic . . . I find it difficult to make peace with the notion that m. 11 marks both the end of the phrase and the end of the section. It would mean cadencing melodically, in the right hand, with A-Bb anticipating the prominent cadence in m. 16 across the divide. Further, it is a most peculiar placement of a compressed phrase, coming as it does at the section's close, unless, perhaps, it is not offered as a phrase ending at all but rather as a way of motivating the repeat of the

Menuett

Example 1 (Schoenberg, op. 25)

Brackets enclose twelve tone rows. Circles indicate melodic cadençes by "stepwise" motion

Example 2
(from Stravinsky
Octet)

Circles indicate
octatonic collection.

Brackets indicate
diatonic collection.

Example 2b Theme

the octatonic scale:

section, i.e., coming as a form of abruption in the hope of finding a more suitable ending the second time through the section. I believe that inference is well-founded because of what is offered us in m. 12 complete, though that measure also serves as a beginning of the opening phrase of the second section (a five measure phrase). What leads me to favor this hearing is not simply the continuity of the phrase-length model that seems consistent with the piece's unfolding, but also the strong parallel of the cadential measures; consider measures 4, 8, and 12 for, among other attributes, their intervallic reduction to stepwise motion.[30]

Shifrin observes the relationship between completion of the twelve-tone row, contextual linear patterns in the music, and the phrase patterns which are articulated by these elements of the structure. What is striking about this quotation (given our current concerns), is its "neutrality" of nomenclature with respect to the Stravinsky-Schoenberg controversy. The technical terms in Shifrin's account might be applied equally well to the music of Stravinsky, to the "Neoclassic" works as well as the "twelve-tone works". That is, Shifrin creates a language for describing and appraising Schoenberg which does not inherently imply a dismissal or devaluation of Stravinsky.

Similarly, in recent scholarship, a technical vocabulary for the elucidation of Stravinsky's music has emerged (or been reinterpreted) which has enabled students to observe similarities between his neoclassic works and Schoenberg's. Unlike the Schoenberg "Menuett", our excerpt from the Stravinsky *Octet* is not based upon twelve-tone principles, but it can be analysed using a nomenclature emphasizing contextually ordered presentations and interrelations of pitch class set. In this case interval relations of the octatonic scale (comprised of alternating half steps and whole steps, see example 2b) are emphasized. An ordering of the octatonic "set" is presented melodically at the beginning of the movement (the theme of the variation set, see example 2 and 2b).

The transformations of this melody are systematically less constrained than a Schoenbergian, twelve-tone melody. For example, unordered fragments of the octatonic scale the interspersed and contrasted with diatonic sets in the passage cited in example 2. But in the first phase of the Schoenberg example, are completion of the transposition of the pitch sequence of the original tune of the movement here in the second trumpet, beginning at rehearsal # 34 and concluding on the concert pitch A in the last measure of the example, signals closure and corresponds to an important phrase and form boundary. In the last several decades, analysts have discussed the relation between "set structure" and phrasing in the music of Schoenberg. Similar properties have been observed in the music of Stravinsky. More and more, similar terms and analytic strategies to those invoked to describe Schoenberg's music have been applied to Stravinsky. We wish to suggest that these recent observations of similarities between the two composers are indicative of a tendency, even in apparently, neutral, "formalist" theoretical nomenclature, to homogenize the achievements of highly esteemed composers by creating or adapting technical language which will reveal the similarities between them.

Another point of comparison, which is revealed by recent methods in music theoretical research, may be mentioned in passing. A careful unfolding of the complete (twelve) pitch class set may be observed in both examples. In twelve-tone music, the unfolding of this set (sometimes referred to as the aggregate) is intrinsic to the definition of the system. In the Stravinsky example, the presentation of the twelve pitch classes occurs by gradual accumulation, not governed by such *a priori*, systematic conventions. Eight of the twelve pitch classes are presented in the first four measures. The other third of the chromatic set is unfolded, one pitch at a time. Each new pitch is emphasized by instrumental doubling or presentation at phrase boundaries. (See the pitches labelled numbers 9 through 12 in example 2.)

Using some of the language and methodology of music theory, we have observed some common attributes in our Stravinsky and Schoenberg examples: recursive presentation of a fixed order of pitch elements, adherence to some classical formal conventions, conventional phraseological patterns, and emphasis on completion of aggregates and ordered sets. By reinterpretation of some of the same concepts previously used to distinguish the achievements of Schoenberg and Stravinsky, contemporary theorists and composers have demonstrated similarities of techniques between the two. Both achievements have been woven into the fabric of contemporary compositional ideals, and concepts have been modified, abandoned or newly invoked to explain or justify the stature of both composers. (Of course, the similarities between the Schoenberg and Stravinsky examples which we cited are *in the music*, in some fundamental senses. But so are a very large number of other similarities and differences that we might consider meaningless or trivial. As the ideals of music theory change — and these ideals are shaped by many "external" considerations, such as public re-evaluation of a composer's work (a complex social process) — the concepts of the discipline are transformed, thus encouraging changing modes of perception and evaluation of the discipline's objects of study. This is not to say that concepts change arbitrarily or that their meaning is unconstrained. Rather, we wish to suggest that the standards of precision in the definition of the concepts of musical composition and theory are not drawn along the same lines as those of different, possibly more compact disciplines.

As we have suggested with respect to prior examples, many of these observations may be restated in the language of essentially contested concepts, as formulated by Gallie and Connolly: the achievements of Stravinsky and Schoenberg are *internally complex*; the *order of importance* or their various features varies in competitive interpretations. Most importantly, the descriptions of their achievements *admit considerable modification in the light of changing circumstances.* Our first set of descriptions of Stravinsky and Schoenberg's music (made during the period of the composition of their neoclassic music) vary greatly from the second set of (recent) descriptions. The terms of discourse have changed, and the conceptual range of persistent terms has been modified. As Connolly suggested, *the rationale for grouping the ingredients* of a cluster concept are dependent on the *normative point* of the these concepts. As the achievements of Schoenberg and Stravinsky became more universally recognized, the terms of discourse and the concepts applied to them are changed or reinterpreted. Theorists naturally attempt to develop an explanatory method which will embrace as many highly valued artifacts as possible. Thus, similarities between the two masters become more evident as the practitioners of the discipline interpret and invent critical terms to reveal these similarities. The normative point of the concepts of musical composition affect the criteria of their use. Thus, the terms for understanding valued achievements (in Toulmin's language, the discipline's "reasons") may be very clearly defined in a particular context, but they may be reinterpreted or discarded to justify different ideals. As we have indicated, the diversity of ideals which may coexist in the discipline of musical composition indicates the impropriety of applying the Kuhnian sense of paradigm to musical composition.

VI. CONTROVERSIES IN MUSIC THEORY

We have seen that numerous systems of compositional ideals may coexist, i.e., that competitions between contemporary compositional systems may end with the acceptance of both competitors in the community of professional musicians. (Or, for that matter, a rejected system may be reevaluated at a later time without disrupting the normal practice of composers; victories of compositional systems are not absolute.) But what of theoretical procedures invoked for the interpretation of prior, universally accepted musical achievements? We might expect more rigid and universally accepted ideals (i.e., a unified discipline) among the community of music theorists. Let us consider, then, an imagined dialogue between music theorists engaged in a controversy over the interpretation of an accepted masterpiece, the opening of the Brahms *Haydn Variations* (see example 3, measures 11—18).

Theorist A: Let's take a look at the middle section of the St. Anthony Chorale.

Theorist B: All right. How do you hear it, A?

Theorist A: Well, I hear it as V, essentially.

Theorist B: Yes, I hear it that way too. But how do you group these eight bars?

Theorist A: I hear it as 6+2.

B: Excuse me?

A: 6+2. I hear the principal structural division occurring between the sixth and seventh measure.

B: But clearly, A, it divides 4+4. The phrasing is symmetrical, the first phrase cadences unambiguously, the melodic contour shifts. Brahms himself marks the midpoint of the passage by changing the orchestration and dynamics precisely there.

A: But no, B — I certainly recognize the things you point to, but, as you yourself suggest, they are superficial. I hear an acceleration of the structural motion in the final two bars of the section. The arrival on the II^6 chord begins the half cadence of bars 17 and 18.

B: I don't hear that at all. The II^6, as you call it, is really a substitution for IV. The "chord" in measure 16 is merely a passing $^6_4$ chord which connects the harmonies of m. 15 and 17, both members of the sub-dominant function class. Measures 15—17 are, essentially, a prolongation of the sub-dominant. At m. 15, the melodic line has

Example 3a

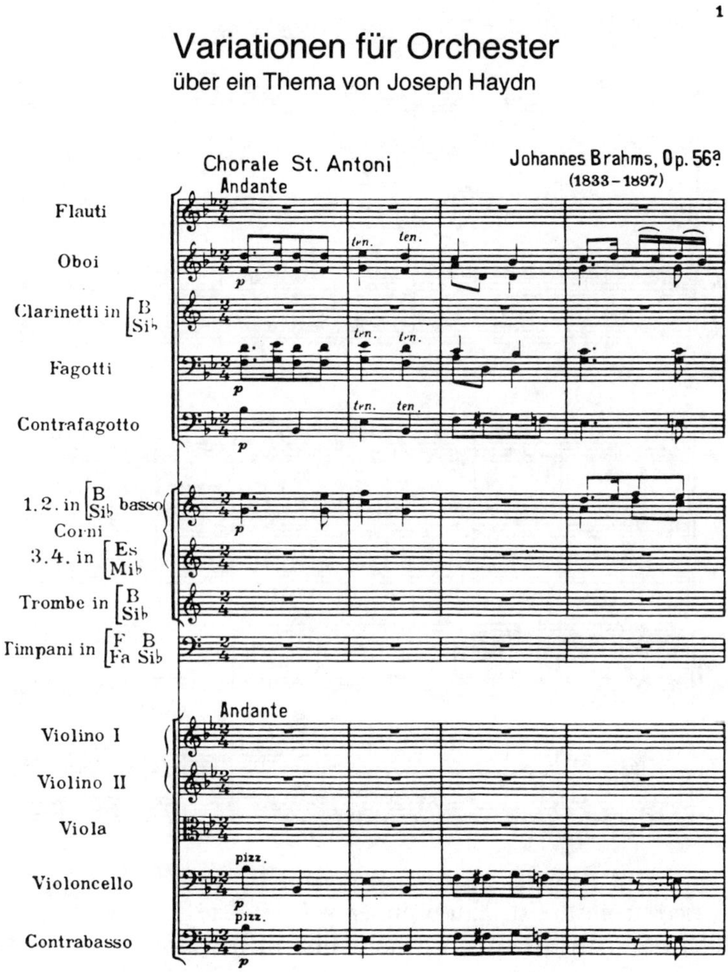
1
Variationen für Orchester
über ein Thema von Joseph Haydn
Chorale St. Antoni
Johannes Brahms, Op. 56a
(1833–1897)
Andante
Flauti
Oboi
Clarinetti in B Sib
Fagotti
Contrafagotto
1.2. in B Sib basso
Corni
3.4. in Es Mib
Trombe in B Sib
Timpani in F B Fa Sib
Andante
Violino I
Violino II
Viola
Violoncello
Contrabasso
ten.
pizz.

reached, from the structural inner voice, up to the upper voice Eb which occurs at the point of initiation of the cadential movement, IV to V.

A: You mean that the II^6 chord in m. 17 has less structural weight than the chord two measures before it?

B: Yes, indeed. The II^6 is essentially a transformation of IV, which is prolonged from measure 15 to 17.

Example 3b

A: But B, if you hear what you say you do, what do you make of the I chord in m. 16?

B: What I chord? There is no real I chord in m. 16.

A: What do you mean?

B: Well, the configuration of notes in m. 16 is, simply, a passing $^{6}_{4}$ chord which connects (by movement in parallel sixths) the stuff in m. 15 and m. 17. In fact, m. 17 is basically IV. The melodic tone C is essentially an accented passing tone and the E natural in the bass is a chromatic passing tone. The structural tone Bb is delayed to avoid parallel fifths between the outer voices (i.e., F moving to Eb in the bass, C moving to Bb in the upper voice).

A: I don't hear that at all — The Bb harmony in m. 16 is indeed functional. In fact, the Bb is reinforced orchestrationally by doubling in the horns, oboe, and bassoon. Bb is the true structural bass tone in m. 16.

B: Let's not get into orchestration now, A. After all, the bass line G-F-Eb is presented not only by cello and bass but also by contrabassoon.

A: OK, but notice the motivic and harmonic correspondences between the first and second sections of the Chorale. The bass line of bars 17 and 18 are exactly the same as the fourth and fifth measures of the opening of the Chorale. In fact, there are two cadences in the first section which move from II to V. Also, in the fourteenth, fifteenth and sixteenth bars of the piece we find an example of the principal motive, the upper neighbor figure, if we connect the most prominent melodic tones of these measures. Surely, you hear the connection between this passage and the first two bars of the piece.

As I said before, I hear Bb as the structural bass tone in the fifteenth and sixteenth bars; so what you call a IV^6 chord in the fifteenth bar is really a second inversion triad formed out of upper neighbors cover an implicit Bb bass.

B: Excuse me, A, but what you claim to hear is outrageous. I do agree with you about one thing, though, and that is that the Eb in m. 15 is an upper neighbor. But this Eb doesn't connect principally to the D of the preceding and following measure, but to the D which occurs in m. 1 and m. 19. This far-reaching neighbor figure binds the whole chorale together as well as reflecting the surface structure, of which you, my friend, are so fond. So, you see, the 4+4 phrasing and the sense of structural accent on the downbeat of the fifteenth measure are quite consistent with the large scale voice leading of the passage.

A: You don't really *hear* those long range connections.

B: That is precisely what I hear.

A: I insist, the structural accent and the principal division occur at the downbeat of the seventeenth bar, where the already familiar cadential pattern, II to V begins.

B: Nonsense. I still don't see how you can hear asymmetrical phrasing in this passage.

A: Well, B, that seems to bring us back to the beginning of our conversation. But don't you think that the phrasing of the first section of the Chorale sets a precedent for the asymmetry which I hear in the second section? By the way, how do you phrase the first section, B?

B: Let's not get into that . . .

A précis of the preceding dialogue may clarify some aspects of the configurations of concepts (the disciplinary "reasons") that each of the two theorists choose to justify their interpretations. B's interpretation of the passage is close to that suggested by Heinrich Schenker in *Der freie Satz*,[34] and B's use of music theoretical nomenclature reveals a Schenkerian perspective. His use of a number of technical terms — "prolongation", "voice leading", "structural tone", "harmony", "grouping" and even "chord" differ from that of theorist A. Both theorists subscribe to a concept of *harmonic prolongation*, i.e., a notion which is invoked to interpret some linear-harmonic patterns in the music as subsidiary to others. These subsidiary patterns are then said to "prolong" the principal harmonic constituents in the passage under consideration. But our two theorists have different criteria for applying the concept *prolongation.* For theorist A, the occurrences of the note pattern D-Eb-D in measures 1—2 and measures 14—16 (where it is elaborated) constitutes an important parallelism in the presentation of a structural motive. Placing high priority on this motivic recurrence (in which D is the principal melodic tone), theorist A argues that the D natural in m. 14 and 16 is (should be considered to be) consonant with the principal bass tone in the passage. Thus, the principal bass tone cannot be Eb — it must be Bb. By this reasoning, the tonic harmony is "prolonged" in this passage. According to theorist B, however, an *a priori* harmonic structure provides the basis for interpreting the passage. Given this assertion, the entire middle section of the *Anthony Chorale* is a *prolongation* of the dominant harmony. The Bb in the bass in m. 14 is not the root of a structurally important harmony, but the terminus of a linear progression in the bass, an element of motion which *prolongs* the more essential dominant harmony. Moreover, the melodic Eb in m. 15 takes priority over the melodic D in m. 14 or 16, because Eb is interpreted as the seventh of the dominant seventh chord being unfolded through the entire middle section. In short, theorist A considers the notion of a "structural motive" to be the guiding criterion for applying the concept of *prolongation.* For theorist B, an *a priori* linear-harmonic structure is the guiding criterion. (Theorist B may also observe a "structural motive" in the passage. In fact, the motive is similar to that of theorist A, but it is instantiated differently in the composition.) Many aspects of the cluster concept, *prolongation*, are similar for both theorists (e.g. notions such as "passing chord", "structural bass tone" even "structural motive"). But "the broader conceptual system in which it [the cluster concept] is implicated"[35] differs.

Our imaginary music theorists share an extensive specialist's vocabulary. They are capable of producing sophisticated "reasons" justifying their activities as music theorists. They invoke professionally-shared metaphors (e.g., "structural weight" or "motion") and problem-solving techniques derived from figured bass theory and the analytic procedures developed by Heinrich Schenker. They appear to be operating within the constraints of something like a shared paradigm. Their conflict, however, is not simply resolved by recourse to the professional standards of the discipline. Even though this competition (for the "right" interpretation of the opening of the Brahms *Haydn Variations*) takes place within the confines of the discipline, this discipline is not sufficiently compact to decide between these alternative interpretations. Our two theorists can call upon different aspects of the definitions of technical terms, as well as different aspects of the musical score being interpreted to justify consistent claims about their readings of the piece. While it may be possible to formulate some aspects of tonal music theory in terms of analytic statements, satisfactory accounts of tonal music are not likely to derive from such formulations. No single theory of tonal music is powerful enough to resolve all disputes of the sort we have presented. We suggest that this is not a shortcoming of music theory and would only be considered to be so if music theory were inaccurately considered an ideally compact discipline.

Notice that both of our imaginary theorists appeal to what they "hear" as an irreducible argument for their interpretations. "Hearing" is discussed as if it were opaque, that is as if it provided irrefutable, transcendant evidence supporting a theoretical claim. But the "hearings" of our theorists are conditioned by a complex network of interpretive strategies — and it is the difference between interpretive strategies which generates the different hearings of our theorists. Thus, the theorist points to his auditory experience to justify his interpretive strategies, and, at the same time, uses his interpretive strategies to inform his auditory experience. The diffuse discipline has not created sufficient modes of justification to escape this tautology — nor do we believe that such modes of justification would be desirable. For alternative modes of interpretation, within the constraints which do exist in the discipline, are both plausible and desirable. This disciplinary open-endedness, whereby multiple interesting and acceptable auditions of the same work may coexist, is often disregarded by music theoriticians, again, in an attempt to treat music theory as an ideally compact discipline.

Another aspect of music theoretical competition may be mentioned in passing. A composer may create artifacts which confirm his own assertions as a theorist. Thus, the formal properties of the twelve-tone system (such as hexachordal combinatoriality) articulated by Schoenberg in his theoretical discussions of twelve-tone music were demonstrated to be musically efficacious by his own compositions. Of course, the compositions must be accepted by the community of professionals comprising the discipline; but the object of study in music is particularly interconnected with aspects of its analysis.

VII. CONCLUSION

Man is a political animal.
Aristotle, *Nichomachean Ethics*

There is a huge body of literature (and it is still growing) which explores the complexities surrounding scientific discovery. One of the few unquestionable premisses of this literature is the notion that describing scientific controversies is problematic. For, as we have suggested, the very terms in which our descriptions are cast can color our interpretation of the discovery. Thus, in the presence of scientific controversies, the details of the description of scientific achievement will itself be controversial. However, what is not controversial is the nature of the achievement. To have discovered the structure of the DNA molecule is to have written the conclusion to a chapter in the history of biology. This is what we mean by the notion that victory in scientific controversies is absolute — it no longer makes sense to contest the achievement. It is important to emphasize that we are speaking only of individual achievements here; we do not mean to suggest that science is anything other than open-ended, that future discoveries will almost certainly force us to revise the substance of a current discovery and perhaps rehabilitate research programs and traditions which seem to have been discredited by our current achievement. Scientific controversies terminate (to the extent that they terminate at all) in achievements whose rational status demands recognition from the scholarly community as a whole. To fail to recognize an achievement of the order of DNA is to brand oneself a "crank": so powerfully unambiguous is the data and reasoning which goes into the articulation of the discovery. This is significantly dissimilar to the case of achievement in music, which is such that it does not demand consensus in the community at large. The fact that theorists, critics, and composers can be working within different traditions, that is, in the context of different "transmits", makes it possible for one theorist, critic, or composer to fail to recognize the achievements of another without *necessarily* branding himself as a "crank" (i.e., in a sense, a non-practitioner of the discipline).

As we have also seen, many controversies within music involve shared elements of technique which are, nonetheless, not necessary and sufficient conditions for the unambiguous resolution of professional disputes. There are many cases in which the resolution of controversy occurs without a definitive victory for either contestant. We have examined (and taken seriously) discourse about music which is blatantly polemical as well as apparently more rational, technically-constrained discourse. We do not believe that the former is somehow pre- or non-paradigmatic, and hence totally incomparable to more "technical" language. The rationality of the concepts invoked by our imaginary music theorists (in section VI) does not confer a special, paradigmatic status on them. Like the language of the Schoenberg-Stravinsky controversy cited above, the technical language of our music theorists has a polemical base. The possibility of rational procedures in music is suggestive of a broad allusion to science idealized — whereby all conflicts might be resolved by practitioners of what we might call "normal music". But this analogy to science and the paradigm language of Kuhn has been shown to be inappropriate.

All of the examples we have presented in the preceding sections are variations of this idea. Adequate discussion of these examples and an adequate adumbration of examples to develop fully our central contention is beyond the scope of this project. We have focussed our attention on the reasons why controversies cannot always be resolved definitively, and by professionally accepted procedures, in music. On the other hand, we wish to make sense of musical controversies in a way that does not lead to despair at the first signs of serious disagreement. We hope to have shown that the mere presence of controversy does not preclude agreement. If the kinds of disagreements we have illustrated are at all resolvable — and this is a formidable condition — it will be the result of examining the structure and origin of the essentially contested issues underlying the controversies.

Our analogy with controversy in science was presented to emphasize that controversy is by no means absent in the "hard sciences". The kinds of impasses which we encounter in the discussion of musical issues should be taken an neither odd nor unique, nor a sign of the discipline's "defective" character.

We have emphasized the way in which musical controversies are embedded in social ones — for example, how a latent analogy to science affects the procedures of music theory, how the techniques of musical analysis are reinterpreted in light of changing public appraisal of musical artifacts, and how composers invoke elaborate metaphors in describing each other's work. In instances of musical controversy, music does not speak for itself. To have a publically articulated disagreement about music, it must be in words. Musical composition, then, spills over into something quite different as soon as we begin to talk about it. It is by no means accidental that the most animated controversialists in twentieth century music were deeply involved in the issues of the day, either as social critics (Schoenberg) or as proponents of innovative techniques in the face of cultural stodginess (Schoenberg again, or Stravinsky, or Varèse, for example, to name only a few).

We have been aiming at provisionally mapping the relationship between composition and: (1) music theory (the cognoscenti) (2) music criticism (the cultural mediators, not necessarily identified with the elite) (3) public appreciation (4) compositional ideologies which may include many things — musical techniques, normative judgements, social ideals, etc. In our discussion, we have suggested that these

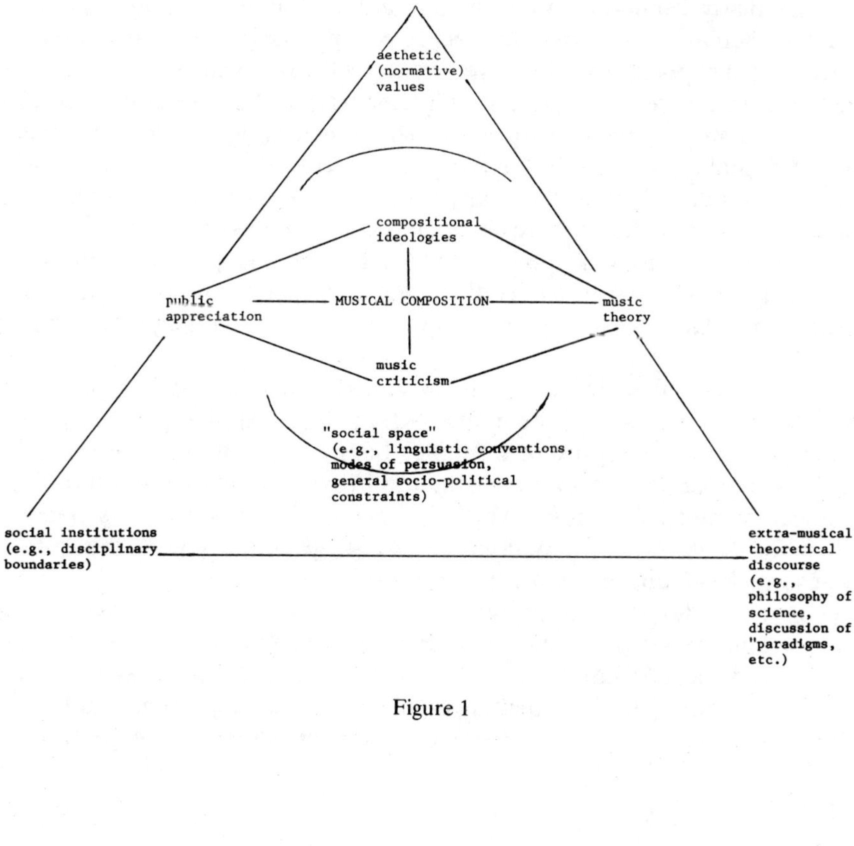

Figure 1

relationships cannot be understood without reference to an even more general body of cultural considerations. Each of these elements has its own history and a history of intersecting with the others. In any given case of musical controversy any one of these variables may become central, affecting any of the others. For the most part these elements require articulation in words; however, the available words will always be part of the current social idiom, and therefore the discussion of these elements will be no less complex than any other cultural debate.

Of course, the constellation of cultural considerations we have mapped above may be considered to be the proper components of music historical studies. Often, however, the range of considerations which inform musical controversies between composers and/or music theorists is forgotten or ignored, as if a "formalist" view of such controversies was adequate.

It is the merit of Gallie and Connolly, in their discussions of essentially contested concepts, to have shown that politics is not one among many other human activities, but that it is a quintessentially human activity, as Aristotle suggested, precisely because humans employ concepts about which we can expect disagreement. Whether or not we are aware of it, these disagreements are the very substance of politics. It should not be surprising that we can designate, for example, the Schoenberg-Stravinsky debate as an episode in the history of *musical politics*. Nor is there anything denigrating in that notion, for where there is culture there is politics. By emphasizing that there is such a thing as musical politics (which is part and parcel of musical culture) we wish to re-assert the idea that there can be persuasion within musical disputes — something which the paradigm language precludes.

As we indicated, the diagram presented above is provisional. This conjectural schema may be improved upon by applying it to many different cases. Our present examples have been chosen to reveal the methodological landscape rather than to provide a full exposition of the subject of musical politics, which is beyond the scope of this paper. Suitably, however, we may conclude by suggesting a few of the questions which our approach might illuminate.

(1) How did the public and professional appraisal of Schoenberg and Stravinsky change from the 1930s to the present? What role did music theoreticians play in this change? What was the role of composers? Influential public performances? When did key theoretical terms and techniques appear which contribute to the

partial homogenization of Stravinsky and Schoenberg's accomplishments? What are the influences (within and outside of music) on the theorists who introduce these techniques and terms?

(2) What are the motivations and intellectual antecedents of the formalist impulse in music theory? To what extent are these motivations and antecedents similar to those which shaped Heinrich Schenker's thinking? (Schenker is a particularly important case, but this question might be applied to numerous other prominent figures.) How does an appreciation of these traditions influence our understanding of the specific analytical techniques and nomenclature which Schenker developed? How does it influence our understanding and appraisal of the different stages of Schenker's work?

(3) To generalize on our fist question, how does a controversial compositional achievement become non-controversial, accepted by professional and public constituencies? More particularly, what is the response of music theoreticians to a compositional achievement which is (on some levels) publically accepted but anomolous, i.e., radically innovative (e.g., Varèse)? Do the terms and procedures of music theory change (perhaps becoming more general) to embrace this newly accredited achievement?

Can we discern any pattern of conditions in the public accredidation of music which *necessitates* recognition by music theorists, leading to the development of analytic procedures to "account" for the music? Can the converse happen? That is, can an innovative achievement in music theory lead to new understanding of a compositional achievement, and, finally, to a revision of public appraisal?

Like some of our earlier inquiries, these provocative questions cannot be answered without much lengthier considerations. Nonetheless, they each suggest the importance of examining all the language of musical description from a broad perspective. For, in a world without musical paradigms, none of the language of musical description is privileged; none is beyond contest and modification under the influence of politics and persuasion.

NOTES

[1] Perhaps the most fully developed application of Kuhn's "paradigm language" to music was made by Steven Haflich in his opening remarks to the colloquium "Analytic

Method: Its Purposes, Premises, Problems, and Polemics", presented at the 1980 national meeting of the Society for Music Theory.

[2] Thomas Kuhn, *The Structure of Scientific Revolutions* (Chicago, 1970); see especially chs. 3—5, pp. 23—51 *etpassim.*

[3] Margaret Masterman, "The Nature of a Paradigm", *Criticism and the Growth of Knowledge*, ed. Imre Lakatos and Alan Musgrave (Cambridge, 1970), pp. 59—90.

[4] See Elisabeth Janeway, "Who is Sylvia: On the Loss of Sexual Paradigms", *Signs*, 5 (1980), pp. 573—589 and T. Hutchinson, *On Revolutions and Progress in Economics* (Cambridge, 1978), *passim.*

[5] See Imre Lakatos, "Falsification and the Methodology of Scientific Research Programs", *Criticism and the Growth of Knowledge*, pp. 91—196.

[6] Stephen Toulmin, *Human Understanding* (Princeton, 1973), vol. **I**, p. 379.

[7] Toulmin, *loc. cit.*

[8] Toulmin, *ibid.*, p. 398.

[9] William Gallie, "Essentially-Contested Concepts", *The Importance of Language*, ed. Max Black (Englewood Cliffs, 1962), p. 125.

[10] Gallie, *loc. cit.*

[11] William Connolly, *The Terms of Political Discourse* (Lexington, Mass., 1974), p. 14.

[12] Connolly, *ibid.*, p. 18.

[13] Connolly, *ibid.*, p. 27—9.

[14] Anne Sayre, *Rosalind Franklin and DNA* (New York, 1975), pp. 112—3.

[15] Our discussion of the sociology of scientific competition is heavily indebted to Sayre's discussion, which is vastly superior to what passes for academic sociology of science. For a classic discussion of the latter see Warren O. Hagstrom, *The Scientific Community* (New York and London, 1965), pp. 69—104; cf. M. J. Mulkay, "The Scientific Research Community," *Science, Technology and Society*, ed. I. Spiegel-Rösing and D. de Solla Price (London and Beverly Hills, 1977), pp. 117—48. The latter contains a useful bibliography.

[16] James D. Watson, *The Double Helix* (New York, 1968), p. 55.

[17] Watson, *ibid.*, p. 56.

[18] Watson, *ibid.*, p. 99.

[19] Watson, *ibid.*, p. 106.

[20] Watson, *ibid.*, pp. 20—1.

[21] Sayre, *ibid.*, p. 207 n. 6.

[22] Our discussion of the work of Hauer will be necessarily elliptical. The theorist-composer developed his ideas in numerous studies (for example, *Zwölftontechnik*: *die Lehre von den Tropen* Vienna, 1926). A brief consideration of Hauer's theory is presented in George Perle, *Twelve-Tone Tonality* (Berkeley and Los Angeles, 1977), p. 1—4.

[23] Thus, attention has been called to Hauer's importance in George Perle, *loc. cit.*, and in George Perle and Paul Lansky, "Twelve-note composition", *New Grove Dictionary of Music and Musicians*, ed. Stanley Sadie, (London, 1980), p. 287. On the other hand, Allen Forte has cast aspersions on Lansky and Perle's reference to Hauer in his review of the theory articles in the New Grove (in *The Musical Quarterly*, LXVIII no. 2, p. 178). Forte suggests that Lansky and Perle have been uncritical of Hauer, failing to point out inconsistencies in his research and presumably over-valuing his achievement.

[24] Gallie, *op. cit.*, p. 125.
[25] In Theodor Adorno, *The Philosophy of Modern Music*, trans. Anne G. Mitchell and Wesley V. Blomster (New York, 1973), *passim.*
[26] In Arnold Schoenberg, *Style and Idea*, ed. Leonard Stein (New York, 1975), p. 482.
[27] *Ibid.*, p. 482—3.
[28] Igor Stravinsky, *Autobiography* (New York, 1936), p. 67.
[29] Arthur Lourie, "Neogothic and Neoclassic", *Modern Music*, V no. 3 (March, 1928), p. 6.
[30] Seymour Shifrin, "*Style and Idea* by Arnold Schoenberg", *Perspectives of New Music*, 14 no. 1, pp. 177—9.
[31] This analytic example is not a quotation from earlier studies, but it is meant to indicate, concisely, how atonal set theory and what we might call "twelve-tone thinking" have effected the discourse on Stravinsky. The approach to Stravinsky suggested here derives in part from Arthur Berger, "Problems of Pitch Organization in Stravinsky", *Perspectives on Schoenberg and Stravinsky*, ed. B. Boretz and E. Cone (Princeton, 1968), pp. 123—55 and Pieter C. Van Den Toorn, "Some Characteristics *of Stravinsky's Diatonic Music*", *Perspectives of New Music*, 15 no. 2, pp. 58—95.
[32] A list of important recent studies would necessitate a lengthy diversion. Writings by numerous prominent theorists such as Babbitt, Benjamin, Forte, Lewin, Hyde, and Perle come to mind.
[33] See, for example, Joseph Straus, "A Principle of Voice Leading in the Music of Stravinsky", *Music Theory Spectrum*, 4 (1982), pp. 106—24.
[34] See Heinrich Schenker, "Supplement: Musical Examples", *Free Composition*, trans. and ed. Ernst Oster (New York, 1979), fig. 42.2.
[35] Connolly, *op. cit.*, p. 14.

AFTERWORD WITH ACKNOWLEDGEMENTS

A few words are necessary here to indicate where these pieces originated, my current opinion of some of the older ones and to acknowledge those debts which are not recognized in the footnotes.

The first five essays originally constituted an anthology which was printed for my students in the University of Bergen (Norway) Philosophy Department Stencil Series. It bears the same title as this volume.

Nothing that has been published since I wrote "Style and Idea in the Later Heidegger" would tempt me to alter my arguments about Heidegger's *Sprachdenken.* I would be prepared to change that assessment sould a direct connection be shown between his later texts and Nazi ideology, i.e., on the basis of his own statements or his unpublished writings. Philosophically, the reception of the later Heidegger outweighs its genesis. Since readers of the most diverse backgrounds both politically and intellectually have acknowledged a great debt to his path of thinking, there is much to be said for the view that his personal beliefs have little to do with its worth.

The Adorno thesis upon which Minder built his case, i.e., the view that "existentialism" was the ideology of a threatened lower middle class in the Great Depression, which ideology was easily absorbed into Nazism, apart from being unspeakably crude reductionism, has received its death blow in the work of Richard F. Hamilton. In his penetrating, *Who Voted for Hitler,* which in 1982 was the first study ever of election returns in the last years of the Weimar Republic, Hamilton has conclusively shown on the basis of data that had been available for fifty years but nevertheless went wholly unexamimed by scholars, that it was the upper middle class that was disproportionately voting for Hitler, whose crucial electoral support came from Protestant *Bauern.* The lower middle class was thus not especially prone to Nazism, which in fact did not have a stable electoral backing but kept managing to attract new voters as the old grew disillusioned.

With respect to the substance of the position which I attribute to Heidegger, i.e., the idea of universal incommensurability the sorts of arguments which Putnam has brought against Kuhn and Feyerabend would appear to hold: how could you articulate it in such a way as to communicate it to someone else. Interestingly Heidegger does not himself argue this point but simply takes it over from Nietzsche.

My critique of Nyíri's conservative interpretation of Wittgenstein is one of the earliest of the pieces collected here. It was written in 1979 and presented at session of the English Austro-German Seminar in 1980, which was devoted to the critique of that thesis. It should be clear from the context in which my arguments appear now that I am more sympathetic to something like a "conservative" reading of Wittgenstein's later philosophy but I still wholly reject Nyíri's picture of him as a conservative ideologue as well as the arguments he advances. In general, I am inclined to see Wittgenstein as making a Promethean effort to transcend categories like "liberal" and "conservative". In fact it is precisely therein that I find his importance for social thought. This essay, like "Wittgenstein, Marx and Sociology" previously appeared in my collection *Essays on Wittgenstein and Weininger* published by Rodopi in Amsterdam in 1985. Both are reprinted here with the kind permission of that publisher.

"Wittgenstein, Marx and Sociology" is an expanded version of a lecture that I gave to the Centre Cavaillès (in its previous reincarnation) at Chantilly in 1985.

"On Edification and Cultural Conversation" was a lecture at the New School for Social Research in February 1983.

"Towards a Wittgensteinian Metaphysics of the Political" was written in Bergen in 1984. It is an effort to clarify my ideas about the significance of Wittgensteinian philosophy for a renewed political philosophy. In fact it is more of a working paper than an article proper.

"Culture, Controversy and the Human Studies" was first published in a Swedish version in the Umeå periodical *Tvärsnitt* in 1986. I include it here as an example of what essential contestability is all about.

"The Politics of Conciliation" was presented at the Divinity School of the University of Chicago in May of 1984 to the symposium "Continental Philosophy — Analytic Philosophy: A New Relationship?". It has the character of a sketch but reflects my longstanding concern with ecumenism in philosophy as well as the role that the history of ideas can play in that effort.

"Discussing Technology" was my contribution to the Swedish Council on Future Oriented Research's symposium "Is the Computer a Tool?" in Sigtuna, Sweden in June 1979. It is reprinted here with the permission of Almqvist & Wiksell who published the proceedings of that conference in 1980 under the same title. Today I would present Ferdinand Ebner in a more subtle light. As I have learned from the seminars I have taught with Walter Methlagl in Innsbruck in 1985 and 1987 Ebner was so extremely suspicious of formalization because he understood its seductions rather than out of a simple hatred for something he did not understand. His confrontation with his sometime friend, Josef Mattias Hauer, inventor of the twelve tone technique for musical composition, is a case in point.

"Socialization Is Creative Because Creativity Is Social" was originally presented to the Nordic Institute for Urban and Regional Planning in February 1986. It is published here for the first time.

"Myth & Certainty" was originally a lecture to the Centre Caviallès symposium on Wittgenstein's *On Certainty* in 1986.The comparison between Wittgenstein and Sorel strikes me as all-too facile today but it nevertheless serves to focus upon some aspects of Wittgenstein's thought which are both often neglected and highly significant for social science.

Permission to cite material from Wittgenstein's *On Certainty* has been granted by his literary executors.

"Self-Knowledge, Naturalism and Certainty" is at once a coming-to-terms with a dissenting perspective of an old and dear friend and the fullest statement of what I find most significant in Wittgenstein's later philosophy. It was presented to the University of Tromsø's symposium on Wittgenstein's *On Certainty* in 1986.

"Psychoanalysis: Art, Science or Literature" was a challenge that Jacques LeRider, editor of the Rouen periodical *Austriaca* threw down to me, i.e., to say something of significance on this topic in fewer than twelve typed pages. It is reprinted here with the gracious permission of the editors.

"Between Enlightenment and Counter-Enlightenment" was written for (but not delivered at) the Centre Cavaillès to correct the faulty picture of Lichtenberg current French debates about the role of philosophy. It is a sampler aimed at tantalizing the reader.

"Tacit Knowledge" and "Paradigms, Politics and Persuasion" are sufficiently described in the preface to this volume as not to require further identification.

In general I have acknowledged my most important intellectual debts in the notes to these papers. However, it would be amiss not to emphasize how important my stay at the University of Bergen in 1983—1984 as Marshall Fellow was both for my understanding of Wittgenstein and for clarification about general features of that philosophy of praxis to which I have adhered since I first studied the *Nichomachean Ethics* of Aristotle with Joe MacDonald at St. Anselm in 1962. To Kjell S. Johannessen and Tore Nordenstam I am most grateful. That goes also for the Norwegian Socrates, Hans Skjervheim.

Finally, this book is dedicated to three friends and colleagues who have unfailingly extended their unconditional support and immense stimulation to me during the period in which these essays were written. It is but a small token of the gratitude that I owe them.

INDEX OF NAMES

Abraham a Santa Clara (Ulrich Megerle), 8—9, 26, 36
Adorno, T. W., 1, 39, 240, 263, 264
Anaximander, 4, 26
Anscombe, G. E. M., 130
Aquinas, St. Thomas, 107
Ariés, Philippe, 134, 145
Aristotle (Aristotelian), xi, xiii, 13, 16, 28, 44, 74, 90, 91, 97, 98, 100, 101, 104, 111, 175, 181, 189, 257
Arnstberg, Karl-Olov, 110—15
Augustine of Hippo, St., 93, 106, 153, 205
Austin, John, 34, 68, 71, 72, 89, 130

Babbitt, Milton, 263
Bachelard, Gaston, 152, 153, 208
Bachtin, Nicholas, 60
Baeumler, Alfred, 19, 35
Bakunin, Mikail, 47
Barkin, Kenneth, 132, 158
Barthes, Roland, 38, 194
Beaton, Mrs., 217
Benjamin, William, 23
Berger, Arthur, 263
Bergson, Henri, 161—65
Bernhardi, Fieldmarshal General von, 124 124
Bernhardt, Josef, 137—38, 141, 145
Bernstein, Basil, 55, 56
Bernstein, Richard, 117
Blessing, Eugen, 39, 58, 145
Blumenthal, Bernhardt, 36
Boltzmann, Ludwig, 50, 66
Bouvresse, Jacques, 132
Boyer, John, 34, 39
Bradley, F. H., 125
Brahms, Johannes, 250—56
Bréhier, Emile, 83
Brinton, Crane, 46, 57
Brod, Max, 32
Brody, Martin, xvi—xvii, 225—263
Bruno, Giordano, 31
Brutus, Marcus, 50
Buber, Martin, 13, 143

Caesar, Julius, 50
Caputo, John, 38
Carlyle, Thomas, 125
Carnap, Rudolf, xii, 12, 36, 88, 121, 151, 189
Cassirer, Ernst, 130
Cavaillès, Jean, 128—29, 130, 132
Cavell, Stanley, 187—88
Chesterton, G. K., 125, 126, 132
Child, Julia, 217
Church, Alonzo, 120
Cioran, E. M., 149, 158
Cohen, Stanley, 145
Collingwood, R. G., xiii, 7, 25, 35, 63, 64, 83, 93, 117, 212, 220, 224
Connolly, William, xv—xvi, 90—1, 92, 95, 96—7, 102, 103, 112—14, 115, 128, 132, 145, 231—33, 249, 260, 262, 263
Copleston, Frederick, 88
Crick, Francis, 234
Crick, Malcolm, 55, 58
Culler, Jonathan, 158

Daester, Uli, 35
Dallago, Carl, 31—2, 49, 138—40, 145
Danbolt, Gunnar, 114—15, 130—31, 132, 158
Danford, John, 61, 74, 79
Darwin, Charles, 139, 150, 168, 180, 189
Däubler, Theodor, 32
Davidson, Donald, xiii
DaVinci, Leonardo, 131

DeMasitre, Joseph, 162
Descartes, René (Cartesian), xiii, 11, 18, 62, 72
Dethloff, Klaus, 158
Dewey, John, xiii, xvi, 68, 75—6, 79, 80—92, 116, 123—127, 132, 154, 156, 158, 160, 171
Diamant, Alfred, 145
Diderot, Denis, xi, 197
Dilthey, Wilhelm, 36, 64, 66
Dostoievski, Fiodor, 44, 48, 59
Dreyfus, Hubert, 133—34, 136, 138, 144
Drury, Maurice, 50, 57, 78, 86
Duchamp, Marcel, 148
Duhem, Pierre, 166
Durkheim, Emile, 63, 64, 65—6, 160

Ebner, Ferdinand, 33, 116, 131, 143, 145, 152, 158, 191, 194, 196, 205, 266
Eckhart, Meister, 26, 28, 38
Ehrenfels, Christian von, 129
Ehrenstein, Alfred, 49
Einstein, Albert, 12
Eisley, Loren, 134, 145, 189
Eliot, T. S., 35
Engels, Friedrich, 65
Ernst, Paul, 49
Eucken, Rudolf, 124

Fann, K. T., 50, 57, 91
Farber, Marvin, 123, 132
Farmer, Fanny, 217
Faulkner, William, 148
Feder, Gottfried, 37
Feinstein, Alvin, 224
Feuerbach, Ludwig, 70
Feyerabend, Paul, 47—8, 57, 116, 264
Ficker, Ludwig von, 30—3, 34, 39, 138—44
Fisher, Seymour, 196
Flanagan, Owen, 38, 188—89
Ford, Henry, 137
Forte, Allen, 262, 263
Foucault, Michel, 160, 171, 184
Francis of Assisi, St., 185
Franklin, Rosalind, 234, 236—37
Franz Joseph I, 51
Frege, Gottlob, 125, 128, 179
Freud, Sigmund, 69, 76, 174—76, 182, 190—96, 208
Frey, Gerhard, 187—88

Gabele, Anton, 4, 8
Gadamer, Hans-Georg, 122, 124
Galilei, Galileo, 228
Gallie, William, xv, 57, 76, 89—90, 92, 96, 110—13, 115, 127—28, 132, 231—33, 240, 249, 260, 262, 263
Garver, Newton, 74—5, 79
Gasman, Daniel, 145
Godwin, William, 47
Goethe, J. W. von, 9, 29, 39
Goldmann, Lucien, 36
Göranzon, Bo, xvii, 188, 24
Gould, Stephen Jay, 158, 189
Grabowsky, Adolf, 50
Greenberg, R. P., 196
Gullers, Peter, 212, 217, 220—21, 224
Gustavsen, Bjørn, 222—23, 224

Habermas, Jürgen, 160, 171, 216, 222
Haeckel, Ernst, 138—39, 143
Haecker, Theodor, 33, 39, 49, 124, 132, 140, 145
Haflich, Steven, 261
Hagstrom, Warren O., 262
Hamann, J. G., 197
Hamilton, Richard, 154—55, 158, 264
Hantsch, Hugo, 9
Hauer, Josef Mattias, 238—40, 262, 266
Hebel, Johann Peter, 1—39
Hegel, G. W. F., 10, 17, 66, 67, 84, 97, 103, 116, 117, 164
Heidegger, Martin, xi, xii, 1—39, 67—8, 79, 80—92, 116, 117, 121, 123, 129, 130, 161, 186, 264—65
Heraclitus, 26
Hertz, Heinrich, 50, 66
Hertzberg, Lars, 210
Hitler, Adolf, 1, 5, 19, 34, 37, 124
Hobbes, Thomas, xi, 69
Holt, Niels R., 145

Horowitz, Vladimir, 215
Hughes, H. Stuart, 159, 161, 171
Hühnerfeld, Paul, 1, 35, 36, 37
Humboldt, W. von, 36
Hume, David, xiii, 125
Husserl, Edmund, xiii, 14, 37, 83, 121, 123, 124, 125, 129
Hutchison, T., 262
Huxley, Aldous, 181, 189
Hyde, Martha, 263

Isocrates, 101

James, Henry, 148
James, William, 14, 56, 68, 110, 118, 131, 160, 164, 168—69, 171, 180, 193, 198, 205
Janeway, Elisabeth, 262
Jaspers, Karl, 95
Jesus of Nazareth, 31
Johannessen, Kjell S., 96, 130, 132, 158, 189, 216, 224, 267
John of the Cross, St. 23, 122
Josefson, Ingela, 212, 217, 219—20, 224
Josephs, Herbert, 209
Joyce, James, 148
Jünger, Ernst, 19

Kadaffi, Moamer, 185
Kafka, Franz, 32
Kann, Robert A., 36
Kant, Immanuel (Kantian, Neo-Kantian), 14, 15, 17, 66, 97, 107, 124, 128, 163, 167, 216
Kaufmann, Walter, 13, 36, 46, 57, 209
Keenan, John, 158
Keller, Evelyn Fox, xviii
Kenny, Anthony, 35, 59, 78, 91
Kierkegaard, Soren, 13, 15, 21, 23—5, 33, 38, 87—8, 124, 140, 203, 204, 205
Klemperer, Klemens von, 44—5, 46, 48, 51, 57, 145
Kolb, David, 35
Krafft-Ebing, Richard von, 190, 194, 195
Kraus, Karl, xi, 31, 32, 49, 51, 52, 138, 140, 190, 205
Kuhn, Thomas S., 16, 18, 20, 42, 47, 80, 81, 87, 118, 134, 135, 145, 164, 198, 226—29, 230, 249, 258, 262, 264
Kurrik, Marie Jaarus, 38

Lackner, Wolfgang, 188
Ladrière, Jean, 39
Laing, R. D., 175, 183
Lakatos, Imre, 47, 133, 144, 229
Lansky, Paul, 262
Latey, Philip, 189
Lavater, Johann Caspar, 201
Lawrence, D. H., 44, 181
Lawrence, T. E., 44
Lecourt, Dominique, xvii
Lehne, Friedrich, 39
Leibniz, G. W. von, 198
Leo XIII, Pope, 224
LeRider, Jacques, 266
Lessing, G. W. von, 38
Lewin, David, 263
Lichtenberg, G. C., 53, 197—210
Lidtke, Vernon, 36
Locke, John, 111
Loos, Adolf, 52
Lourie, Arthur, 241, 263
Loyola, St. Ignatius, 21, 22—3, 24, 25, 27, 38, 122
Lubasz, Heinz R. (Harry), 151, 158
Lueger, Karl, 34
Luther, Martin, 2—3, 9, 10, 124

Mach, Ernst, 202—3, 205
Machiavelli, Nicoló, xi, 69
MacIntyre, Alasdair, xv, 43, 45, 57, 65, 79, 88, 89, 91, 174, 178, 189, 193—94, 196
Mairs, Nancy, 184, 185, 189
Malcolm, Janet, 196
Malcolm, Norman, 187, 189
Mallarmé, Stephane, 148
Mannheim, Karl, 160
Marx, Karl, 56, 58, 59—79, 93, 97, 99, 103, 116, 116, 221, 224

Masson, J. Mousaieff, 193, 194—95, 196
Mauthner, Franz, 209
Mauthner, Fritz, 38
Mayr, Ernst, 158
McDonald, Joseph B., 267
McGrath, William, 47, 57
McGuinness, Brian, 175, 189, 209
McTaggart, John McTaggart Elis, 125
Megerle, Ulrich (see Abraham a Santa Clara)
Meinong, Alexius von, 129
Meløe, Jacob, 130
Mendel, Gregor, 150, 199
Merzekowski, Dimitri, 49
Mesmer, F. A., 201—2
Methlagl, Walter, 39, 144, 266
Meyer, Siri, 130
Mill, John Stuart, 111
Millet, Jean-François, 32
Minder, Robert, xi, xii, 1—39, 264
Mitzman, Arthur, 132
Moeller van den Bruck, Arthur, 49
Mohler, Armin, 44
Montaigne, M. E. de, 53
Moore, G. E., 36, 116, 125, 126
Mosse, George, 35
Mulkay, M. J., 262

Naegeli, C. W. von, 147, 150
Neumann, Sigmund, 57
Neurath, Otto, xiii, 95, 116, 122, 129
Newman, John Henry Cardinal, 124
Newton Isaac, 18
Nietzsche, Friedrich, 17—18, 19, 32, 69, 76, 84, 87, 96, 119, 124, 205, 207—9, 265
Nordenstam, Tore, xvi, 130, 158, 267
Nyíri, J. C., xi, 40—58, 59, 68, 70, 79, 95, 265

Oakeshott, Michael, 90
Olsson, Gunnar, xi, 148, 158

Parmenides, 26
Pascal, Fania, 60, 78
Pascal, Roy, 60, 70
Passmore, John, xi, 71, 79, 132
Paul, St., 128
Pauling, Linus, 235—36
Peirce, C. S., 113
Perby, Maja-Lisa, 217, 218—19, 220, 224
Perle, George, 262, 263
Philips, Derek, 61
Picasso, Pablo, 149
Pitkin, H. F., 54, 56, 58, 61, 68—9, 71, 73—4, 76, 79, 97
Plato (Platonism), 11, 12, 13, 16—18, 20, 21—2, 26, 29, 34, 37, 87, 122, 163, 224
Podach, E. F., 209
Polanyi, Michael, 214, 224
Popkin, Richard, 88, 91
Popper, K. R., 36, 88, 91, 131, 153, 158, 162, 190, 193—94
Promies, Wolfgang, 209
Protagoras, 101
Proust, Marcel, 126
Putnam, Hilary, 185, 264—65

Quine, W. V. O., xii, xiii, 127, 166, 181, 200

Ramirez, Santiago, xvii, 132
Ramsey, Frank Plumpton, 60, 82
Rawls, John, 23, 216
Rayez, André, 38
Rembrandt van Rijn, H., 128
Rhees, Rush, 52, 57
Rorty, Richard, xiii, 68, 79, 80—92, 118, 119, 129, 131
Rosenberg, Alfred, 37
Rossi-Landi, Ferrucio, 60, 67, 68, 69, 70, 79, 95
Rossiter, Clinton, 43—4, 57
Rousseau, Henri (Le Douannier), xii
Rousseau, Jean-Jacques, 98, 111, 197, 207
Rubinstein, David, 59—79, 95
Ruskin, John, 125
Russell, Bertrand, xiii, 15, 52, 54, 72, 83, 116, 121, 125, 126, 142, 179
Ryan, Alan, 61

Ryle, Gilbert, 68, 71

Saussure, F. de, 151, 158
Sayre, Anne, 234, 236—37, 262
Schäfer, Wilhelm, 3, 35
Scheler, Max, 123—27, 129, 130, 132
Schenker, Heinrich, 254—55, 261, 263
Schlick, Moritz, 47
Schleirmacher, Friedrich, 17
Schmitt, Carl, 35
Schnitzler, Arthur, 190, 192—96
Schoenberg, Arnold, 238—69, 263
Schopenhauer, Arthur, 19
Schubert, Franz, 21
Schumann, Robert, 48
Searle, John, 121
Segantini, Giovanni, 32
Seigfried, Hans, 39
Shaw, G. B., 125
Shifrin, Seymour, 243—47, 263
Simmel, Georg, 126
Simon, Herbert, 133
Skinner, B. F., 173, 203
Skirbekk, Gunnar, 132
Skjervheim, 130, 267
Smith, Adam, 125
Smith, Barry, 129, 132
Spencer, Herbert, 125, 138, 143
Socrates, 31, 53, 177, 223, 257
Sokel, Walter, 36
Sombart, Wener, 125, 126
Sorel, Georges, 159—71, 266
Spengler, Oswald, 40, 59
Spinoza, 208
Sraffa, Piero, 50, 60
Steinbüchel, Theodor, 39, 145
Steiner, George, 5, 35
Stern, J. P., 34, 37, 39, 46
Stevenson, C. L., 112
Stieg, Gerald, 39, 144
Straus, Joseph, 263
Stravinsky, Igor, 238—63
Strawson, Peter, 214, 224
Strindberg, August, 172, 187, 188, 189
Stumpf, Carl, 129
Szasz, Thomas, 175, 183

Talleyrand-Périgord, Charles-Marie, 1115
Thébault, André, 38
Thomas à Kempis, 122
Thompson, Edward, 43, 57
Thompson, J. B., 186, 189
Thucydides, 13
Tilghman, Ben, xii—xiii
Tolstoi, Leo, 59
Tönnies, Ferdinand, 126
Toulmin, Stephen, xvi, xvii, 71, 118, 132, 145, 158, 188, 200, 229—31, 233, 235, 249, 262
Trakl, Georg, 4, 25, 26—7, 30, 32, 79
Traversa, Gregor, 158
Tsu, Chuang, 150, 158

Urban, Bernd, 196

Van Den Doorn, Peter C., 263
Varèse, Edgard, 259, 261
Vialleneix, Nelly, 38
Viklund, Birger, 224
Villiers de l'Isle Adam, M., 118
Voltaire, 197

Wagner, Richard, 3, 9, 19
Waley, Arthur, 158
Watson, James, 233—38, 262
Weber, Max, 63, 64, 65—6, 126, 155, 160
Webern, Anton von, 242
Weininger, Otto, 59, 191, 205
Weizenbaum, 133—35, 136, 138, 144
Wells, H. G., 117
Whitman, Walt, 32
Wilde, Oscar, 125
Wider, Thornton, vi
Wilkins, Maurice, 234, 236
Wilson, E. O., 136
Winch, Peter, xvi
Windelband, Wilhelm, 88
Winkler, Ken, 35
Wittgenstein, Ludwig, xi—xiv, xvii, 7, 10, 14—16, 28—9, 36, 40—108, 110, 113, 121, 129, 130, 131, 143,

145, 147, 149, 152—53, 156—57, 158, 159—89, 191, 194, 196, 199, 202, 203—9, 214—17, 220, 223, 224, 265, 266
Wolf, Norbert, R., 35
Wright, G. H. von, 40, 57, 209

Yates, Frances, 88
Young, Jock, 145

Zeppa, Melissa, 158
Ziman, John, 189
Zweig, Stefan, 51, 57

BOSTON STUDIES IN THE PHILOSOPHY OF SCIENCE

Editors:

ROBERT S. COHEN and MARX W. WARTOFSKY
(Boston University)

1. Marx W. Wartofsky (ed.), *Proceedings of the Boston Colloquium for the Philosophy of Science 1961–1962*. 1963.
2. Robert S. Cohen and Marx W. Wartofsky (eds.), *In Honor of Philipp Frank*. 1965.
3. Robert S. Cohen and Marx W. Wartofsky (eds.), *Proceedings of the Boston Colloquium for the Philosophy of Science 1964–1966. In Memory of Norwood Russell Hanson*. 1967.
4. Robert S. Cohen and Marx W. Wartofsky (eds.), *Proceedings of the Boston Colloquium for the Philosophy of Science 1966–1968*. 1969.
5. Robert S. Cohen and Marx W. Wartofsky (eds.), *Proceedings of the Boston Colloquium for the Philosophy of Science 1966–1968*. 1969.
6. Robert S. Cohen and Raymond J. Seeger (eds.), *Ernst Mach: Physicist and Philosopher*. 1970.
7. Milic Čapek, *Bergson and Modern Physics*. 1971.
8. Roger C. Buck and Robert S. Cohen (eds.), *PSA 1970. In Memory of Rudolf Carnap*. 1971.
9. A. A. Zinov'ev, *Foundations of the Logical Theory of Scientific Knowledge (Complex Logic)*. (Revised and enlarged English edition with an appendix by G. A. Smirnov, E. A. Sidorenka, A. M. Fedina, and L. A. Bobrova). 1973.
10. Ladislav Tondl, *Scientific Procedures*. 1973.
11. R. J. Seeger and Robert S. Cohen (eds.), *Philosophical Foundations of Science*. 1974.
12. Adolf Grünbaum, *Philosophical Problems of Space and Time*. (Second, enlarged edition). 1973.
13. Robert S. Cohen and Marx W. Wartofsky (eds.), *Logical and Epistemological Studies in Contemporary Physics*. 1973.
14. Robert S. Cohen and Marx W. Wartofsky (eds.), *Methodological and Historical Essays in the Natural and Social Sciences. Proceedings of the Boston Colloquium for the Philosophy of Science 1969–1972*. 1974.
15. Robert S. Cohen, J. J. Stachel, and Marx W. Wartofsky (eds.), *For Dirk Struik. Scientific, Historical and Political Essays in Honor of Dirk Struik*. 1974.
16. Norman Geschwind, *Selected Papers on Language and the Brain*. 1974
17. B. G. Kuznetsov, *Reason and Being: Studies in Classical Rationalism and Non-Classical Science*. 1987
18. Peter Mittelstaedt, *Philosophical Problems of Modern Physics*. 1976
19. Henry Mehlberg, *Time, Causality, and the Quantum Theory* (2 vols.). 1980.

20. Kenneth F. Schaffner and Robert S. Cohen (eds.), *Proceedings of the 1972 Biennial Meeting, Philosophy of Science Association*. 1974
21. R. S. Cohen and J. J. Stachel (eds.), *Selected Papers of Léon Rosenfeld*. 1978.
22. Milic Čapek (ed.), *The Concepts of Space and Time. Their Structure and Their Development*. 1976.
23. Marjorie Grene, *The Understanding of Nature, Essays in the Philosophy of Biology*. 1974.
24. Don Ihde, *Technics and Praxis. A Philosophy of Technology*. 1978.
25. Jaakko Hintikka and Unto Remes, *The Method of Analysis. Its Geometrical Origin and Its General Significance*. 1974.
26. John Emery Murdoch and Edith Dudley Sylla, *The Cultural Context of Medieval Learning*. 1975.
27. Marjorie Grene and Everett Mendelsohn (eds.), *Topics in the Philosophy of Biology*. 1976.
28. Joseph Agassi, *Science in Flux*. 1975.
29. Jerzy J. Wiatr (ed.), *Polish Essays in the Methodology of the Social Sciences*. 1979.
30. Peter Janich, *Protophysics of Time*. 1985.
31. Robert S. Cohen and Marx W. Wartofsky (eds.), *Language, Logic and Method*. 1983.
32. R. S. Cohen, C. A. Hooker, A. C. Michalos, and J. W. van Evra (eds.), *PSA 1974: Proceedings of the 1974 Biennial Meeting of the Philosophy of Science Association*. 1976.
33. Gerald Holton and William Blanpied (eds.), *Science and Its Public: The Changing Relationship*. 1976.
34. Mirko D. Grmek (ed.), *On Scientific Discovery*. 1980.
35. Stefan Amsterdamski, *Between Experience and Metaphysics. Philosophical Problems of the Evolution of Science*. 1975.
36. Mihailo Marković and Gajo Petrović (eds.), *Praxis, Yugoslav Essays in the Philosophy and Methodology of the Social Sciences*. 1979.
37. Hermann von Helmholtz, *Epistemological Writings. The Paul Hertz/Moritz Schlick Centenary Edition of 1921 with Notes and Commentary by the Editors*. (Newly translated by Malcolm F. Lowe. Edited, with an Introduction and Bibliography, by Robert S. Cohen and Yehuda Elkana). 1977.
38. R. M. Martin, *Pragmatics, Truth, and Language*. 1979.
39. R. S. Cohen, P. K. Feyerabend, and M. W. Wartofsky (eds.), *Essays in Memory of Imre Lakatos*. 1976.
40. B. M. Kedrov and V. Sadovsky. *Current Soviet Studies in the Philosophy of Science*. Forthcoming.
41. M. Raphael, *Theorie des Geistigen Schaffens auf Marxistischer Grundlage*. Forthcoming.
42. Humberto R. Maturana and Francisco J. Varela, *Autopoiesis and Cognition. The Realization of the Living*. 1980.
43. A. Kasher (ed.), *Language in Focus: Foundations, Methods and Systems. Essays Dedicated to Yehoshua Bar-Hillel*. 1976.
44. Trân Duc Thao, *Investigations into the Origin of Language and Consciousness*. (Translated by Daniel J. Herman and Robert L. Armstrong; edited by Carolyn

R. Fawcett and Robert S. Cohen). 1984.
45. A. Ishimoto (ed.), *Japanese Studies in the History and Philosophy of Science.*
46. Peter L. Kapitza, *Experiment, Theory, Practice.* 1980.
47. Maria L. Dalla Chiara (ed.), *Italian Studies in the Philosophy of Science.* 1980.
48. Marx W. Wartofsky, *Models: Representation and the Scientific Understanding.* 1979.
49. Trân Duc Thao, *Phenomenology and Dialectical Materialism.* 1985.
50. Yehuda Fried and Joseph Agassi, *Paranoia: A Study in Diagnosis.* 1976.
51. Kurt H. Wolff, *Surrender and Catch: Experience and Inquiry Today.* 1976.
52. Karel Kosik, *Dialectics of the Concrete.* 1976.
53. Nelson Goodman, *The Structure of Appearance.* (Third edition). 1977.
54. Herbert A. Simon, *Models of Discovery and Other Topics in the Methods of Science.* 1977.
55. Morris Lazerowitz, *The Language of Philosophy. Freud and Wittgenstein.* 1977.
56. Thomas Nickles (ed.), *Scientific Discovery, Logic, and Rationality.* 1980.
57. Joseph Margolis, *Persons and Minds. The Prospects of Nonreductive Materialism.* 1977.
58. G. Radnitzky and G. Andersson (eds.), *Progress and Rationality in Science,* 1978.
59. Gerard Radnitzky and Gunnar Andersson (eds.), *The Structure and Development of Science.* 1979.
60. Thomas Nickles (ed.), *Scientific Discovery: Case Studies.* 1980.
61. Maurice A. Finocchiaro, *Galileo and the Art of Reasoning.* 1980.
62. William A. Wallace, *Prelude to Galileo.* 1981.
63. Friedrich Rapp, *Analytical Philosophy of Technology.* 1981.
64. Robert S. Cohen and Marx W. Wartofsky (eds.), *Hegel and the Sciences.* 1984.
65. Joseph Agassi, *Science and Society.* 1981.
66. Ladislav Tondl, *Problems of Semantics.* 1981.
67. Joseph Agassi and Robert S. Cohen (eds.), *Scientific Philosophy Today.* 1982.
68. Wuadysuaw Krajewski (ed.), *Polish Essays in the Philosophy of the Natural Sciences.* 1982.
69. James H. Fetzer, *Scientific Knowledge.* 1981.
70. Stephen Grossberg, *Studies of Mind and Brain.* 1982.
71. Robert S. Cohen and Marx W. Wartofsky (eds.), *Epistemology, Methodology, and the Social Sciences.* 1983.
72. Karel Berka, *Measurement.* 1983.
73. G. L. Pandit, *The Structure and Growth of Scientific Knowledge.* 1983.
74. A. A. Zinov'ev, *Logical Physics.* 1983.
75. Gilles-Gaston Granger, *Formal Thought and the Sciences of Man.* 1983.
76. R. S. Cohen and L. Laudan (eds.), *Physics, Philosophy and Psychoanalysis.* 1983.
77. G. Böhme et al., *Finalization in Science,* ed. by W. Schäfer. 1983.
78. D. Shapere, *Reason and the Search for Knowledge.* 1983.
79. G. Andersson, *Rationality in Science and Politics.* 1984.
80. P. T. Durbin and F. Rapp, *Philosophy and Technology.* 1984.
81. M. Marković, *Dialectical Theory of Meaning.* 1984.

82. R. S. Cohen and M. W. Wartofsky, *Physical Sciences and History of Physics*. 1984.
83. E. Meyerson, *The Relativistic Deduction*. 1985.
84. R. S. Cohen and M. W. Wartofsky, *Methodology, Metaphysics and the History of Sciences*. 1984.
85. György Tamás, *The Logic of Categories*. 1985.
86. Sergio L. de C. Fernandes, *Foundations of Objective Knowledge*. 1985.
87. Robert S. Cohen and Thomas Schnelle (eds.), *Cognition and Fact*. 1985.
88. Gideon Freudenthal, *Atom and Individual in the Age of Newton*. 1985.
89. A. Donagan, A. N. Perovich, Jr., and M. V. Wedin (eds.), *Human Nature and Natural Knowledge*. 1985.
90. C. Mitcham and A. Huning (eds.), *Philosophy and Technology II*. 1986.
91. M. Grene and D. Nails (eds.), *Spinoza and the Sciences*. 1986.
92. S. P. Turner, *The Search for a Methodology of Social Science*. 1986.
93. I. C. Jarvie, *Thinking about Society: Theory and Practice*. 1986.
94. Edna Ullmann-Margalit (ed.), *The Kaleidoscope of Science*. 1986.
95. Edna Ullmann-Margalit (ed.), *The Prism of Science*. 1986.
96. G. Markus, *Language and Production*. 1986.
97. F. Amrine, F. J. Zucker, and H. Wheeler (eds.), *Goethe and the Sciences: A Reappraisal*. 1987.
98. Joseph C. Pitt and Marcella Pera (eds.), *Rational Changes in Science*. 1987.
99. O. Costa de Beauregard, *Time, the Physical Magnitude*. 1987.
100. Abner Shimony and Debra Nails (eds.), *Naturalistic Epistemology: A Symposium of Two Decades*. 1987.
101. Nathan Rotenstreich, *Time and Meaning in History*. 1987.
102. David B. Zilberman (ed.), *The Birth of Meaning in Hindu Thought*. 1987.
103. Thomas F. Glick (ed.), *The Comparative Reception of Relativity*. 1987.
104. Zellig Harris *et al.*, *The Form of Information in Science*. 1987
105. Frederick Burwick, *Approaches to Organic Form: Permutations in Science and Culture*. 1987.
106. M. Almási, *Philosophy of Appearances*. Forthcoming.
107. S. Hook, W. L. O'Neill, and R. O'Toole, *Philosophy, History and Social Action. Essays in Honor of Lewis Feuer*. 1988.
108. I. Hronszky, M. Fehér, and B. Dajka (eds.), *Scientific Knowledge Socialized. Selected Proceedings of the Fifth Joint International Conference on History and Philosophy of Science Organized by the IUHPS, Veszprém, 1984*. Forthcoming.
109. P. Tillers and E. D. Green (eds.), *Probability and Inference in the Law of Evidence. The Uses and Limits of Bayesianism*. 1988.
110. E. Ullmann-Margalit (ed.), *Science in Reflection. The Israel Colloquium: Studies in History, Philosophy, and Sociology of Science*. 1988.
111. K. Gavroglu, Y. Goudaroulis, and P. Nicolacopoulos (eds.), *Imre Lakatos and Theories of Scientific Change*. 1989.
112. Barry Glassner and Jonathan D. Moreno (eds.), *The Qualitative-Quantitative Distinction in the Social Sciences*. 1989.